The Right Seat

An Introduction to Flying for Pilot's Companions and Would-be Pilots

by Avram Goldstein, CFI, CFII

Published by
Airguide Publications, Inc.
1207 Pine Ave., P.O. Box 1288, Long Beach CA 90801

CONTENTS

0

INTRODUCTION

Karen's new boyfriend Walt is an experienced pilot. Karen wants to be a good companion and share Walt's enthusiasm for flying, but the truth is she's terrified of it. Every time the newspapers report an accident to a small plane, Karen's fear gets worse. Curiously enough, newspaper accounts of automobile accidents don't affect her at all, even though she has friends and acquaintances who have been injured in automobiles. Her fear of flying seems to be irrational, and it is beginning to come between her and Walt.

John has always wanted to fly. Down at our local airport he often hangs around, angling for an invitation to go up. Not long ago I took him along on a business flight to a nearby city. As we were cruising along, I let him take the controls and fly the airplane. I couldn't help enjoying the thrill he was getting as he guided the airplane through the skies. Together we watched the hills and fields and towns slip by beneath us. John loves flying but he's not sure yet if he wants to go for his private pilot's license. Some day maybe he will. Meanwhile, he enjoys learning a little more about aviation every time he flies as a pilot's companion.

Margaret's friend Susan just earned her pilot's license. Susan's ambition is to fly for the airlines some day. Both women are college students, and the cost of learning to fly is out of the question for Margaret right now. Nevertheless, she would like to learn as much as she can, to be able to help Susan when they fly together. Also, she would like to be able to join in discussions about flying with Susan and her pilot friends.

Barbara is a busy career woman. Her husband Don owns an airplane for their joint consulting business, and he flies a great deal. Barbara goes along, when she has the time, but she is tense and uncomfortable all the way. "Suppose Don had a heart attack," she thinks, "what would happen to us? I guess we would crash and both of us would be killed." Barbara was surprised to learn that with a little knowledge and practical preparation she would be able to handle the airplane, that she could not only save her own life but also have a good chance of getting Don down to earth and to emergency medical care quickly, should her nightmare ever come true. In fact—as she found out—this has happened successfully on several occasions.

I wrote this book for Karen, John, Margaret, Barbara—and you. My purpose is to teach you enough about small airplanes so that you can:

- be more at ease as a passenger;

- understand more about what is happening at every phase of a flight;

- be of real assistance to the pilot in various ways;

- be able to share the joy of flying more fully with a friend who is a pilot;

- know enough to converse intelligently with pilots;

- be capable of managing the airplane and landing it safely, should the pilot become incapacitated.

- make a better informed decision about whether or not to take flying lessons yourself.

The book is called *The Right Seat* because it is a time-honored tradition in aviation that the pilot sits in the left seat, while a passenger sits on the right. The Wright brothers' machine was flown by a single person lying prone in a central position on the airframe. Early two-seater airplanes had both seats in line, usually with the passenger in front, the pilot in the rear. The first side-by-side seating was introduced in the United States, where automobiles were driven from the left seat, so it was natural to adopt the same convention.

Actually, nearly all aircraft can be flown equally well from either seat, since they have a complete set of dual controls. It is true that the principal flight instruments are usually centered in front of the pilot on the left side of the instrument panel, but these instruments are often duplicated on the right; and anyway, they can always be seen from the right seat. I am going to assume you occupy the right seat when you go flying, and I am going to write as though you have never actually handled an airplane before. If, in fact, you have already assisted a pilot friend in one way or another, you will be able to progress all the faster in becoming a competent pilot's companion, and maybe eventually becoming a pilot yourself.

Chapter 1 will acquaint you with a variety of general aviation aircraft. "General aviation" means all kinds of flying except for air line companies and the military. We shall examine some similarities between different airplanes, and some differences. The main point of the chapter is to familiarize you with the working parts of the airplane, and with its instruments, and to give you a good basic understanding of how it flies.

Chapter 2 will explain how pilots know where they are and where they are going. It will teach you how to use an aeronautical chart and how to use radio navigation.

Chapter 3 is about using the radiotelephone (called a communications or "COM" radio) to talk to controllers and others on the ground and to receive useful information and instructions from them.

Chapter 4 presents a menu of possible ways to be helpful to a pilot as you ride along in the right seat. As the captain is in command on a ship, so is the pilot in absolute command in an airplane. But if the pilot is willing, there are many useful things you can do, such as handling the communications, keeping watch outside for traffic (other airplanes), plotting and following the course on a map, getting weather information by radiotelephone, even flying the airplane while the pilot takes a break.

Chapter 5 explains, in detail, how to handle an airplane and bring it to a safe and speedy landing if the pilot should become disabled.

This has actually been done successfully on several occasions by non-pilots with help from the ground by radiotelephone, and even without any preparation at all. With some forethought and preparation it becomes a relatively simple procedure. My hope is that the chapter will inspire you to seek actual "pinch-hitter" instruction from a qualified flight instructor. This specialized instruction is focused on just what you need to do to bring an aircraft down to a safe landing. Four or five hours is all it takes, and being prepared is like an insurance policy; it is well worthwhile even though one hopes it will never be put to use.

At the end of every chapter is a problem set, to let you try out your knowledge. And the Appendix contains answers and discussion for all the problem sets.

You may be surprised at my use of "they", "their", and "them" as singular pronouns. All aspects of aviation are of interest to men and women alike, and all the opportunities for professional as well as amateur flying are equally available to both sexes. Most authors who write "he", "his, and "him" have a mental image of a man, and the reader gets the same impression. To say that "he" really means "he or she" doesn't solve the problem of the subtle sexist influence of all-male personal pronouns. But I find "he or she" clumsy, especially when repeated again and again. The solution, which I have adopted throughout this book, is in every dictionary—even the venerable Oxford English Dictionary: "They" and "their" and "them" can be used for the singular, meaning "he or she", "his or her", "him or her", when the sex is unspecified. And although it sounds odd at first, one soon gets used to it. After all, the same thing happened long ago with "thou", "thy", and "thee"; "you", which had been permissible only as a plural form, came into use in the singular, and now we think nothing of it.

Alice Schwartz, Claire Greene, and Ann Elsbach were kind enough to read drafts of the manuscript and to make numerous thoughtful comments. I am grateful to them. The book is much improved as a result of their labors, but of course they bear no responsibility for errors or murky passages that may remain.

1

THE AIRPLANE:
A USER-FRIENDLY INTRODUCTION

A WALK AROUND THE AIRPORT

To understand how an airplane flies, let's first examine how it is built. Which parts are essential, which are not? I am going to illustrate by talking about the common two- to six-place general aviation aircraft—the typical Cessna, Piper, Beech, or Mooney product. But most of what I say is just as applicable to corporate jets, wide-body jumbo jet air carriers, and military aircraft. The amazing variety of airplanes lends spice and interest to just knowing about all the different makes and models. Each differs in some manner from every other, yet all have the few common features that are really necessary for flight. Let's look at some of the similarities and differences.

Walk around the general aviation part of any airport, where the small planes are tied down. Inspect the various kinds, and see how many differences you can identify. But take care! Keep your distance, watch out for moving aircraft, and give propellers a wide berth (they might be fired up unexpectedly). Be especially careful to hold children very firmly by the hand at all times.

Most general aviation airplanes, you will see, are "singles"—they have one engine driving one propeller up front (Fig. 1.1). But there are also quite a few "twins", with one engine mounted on each side (Fig. 1.2). You may even find some odd ones—a single pusher prop facing backward, or two engines in line with one pulling and one pushing.

Propellers come in two main varieties—two-bladed and three-bladed—shown in Figs. 1.1 and 1.2, respectively. Some have the blades mounted rigidly to the central hub; these are called "fixed-pitch". Others, called "variable-pitch", have a more sophisticated arrangement, with each blade mounted to the hub in such a manner as to permit a slight rotation to change the pitch (the angle at which the spinning prop cuts through the air). Stand safely back and look at the clever construction of a prop blade, shown well in Fig. 1.1. As you stand in front, the typical prop will spin counterclockwise. The leading edge of a blade, which would cut downward through the air on your left (on the right as viewed from inside the cabin), is thicker and blunter than the trailing edge. Most important, there is a slight twist, just enough for the blade to drive a lot of air backward as it spins. When air is driven backward, an opposite reaction drives the aircraft forward, much as pushing water back with an oar drives a rowboat forward.

Fig. 1.1. A typical four-place high-wing airplane with tricycle landing gear. The wheel covers (called "fairings") are to streamline the airflow and reduce the drag; they are found only on fixed (non-retractable) gear. The twist of the propeller blades is clearly visible in this photo.

Usually the landing gear is of the tricycle type—two main gear under the cabin and one nose-gear out in front, as in Figs. 1.1 and 1.2. In tail-wheel aircraft, however, there are two main gear and a small wheel or skid on the tail (Fig. 1.3). This was the common type in the early days of aviation, hence the term "conventional gear". They are also known, affectionately, as "taildraggers". In the typical small general aviation airplane the gear is fixed—it cannot be retracted. In the more sophisticated airplanes, however (as in Fig. 1.2), the gear can be tucked up into the belly. The advantage of retractable gear is that the air resistance ("drag") caused by the wheels hanging down in the airstream can be reduced, thus improving the efficiency of flight, to get more speed with less fuel consumption.

Fig. 1.2. A sophisticated eight-place retractable-gear light twin. Both main gear fold up toward the middle, the nose gear folds back, and the gear doors then close to make a smooth surface covering the gear wells. Note the three-bladed variable-pitch props. The leading edges of the wings, horizontal stabilizer, and vertical stabilizer appear dark because they are covered with rubber "boots" that can inflate to break off adherent ice. The dark stripe on each prop blade is an anti-icing heating unit.

The body of the airplane—the fuselage—is the box-like enclosure, which is streamlined for minimum air resistance, and which usually provides a sheltered cabin. It is analogous to the chassis of an automobile. In the simplest trainers, the cabin accommodates only two people, generally side by side, but sometimes in tandem. Most general aviation aircraft have four seats—two in front, two in the rear—and often also a small baggage compartment. In the early days, however, there were no cabins; pilots and passengers wore helmets and goggles and got wet when it rained (Fig. 1.4). And in the first airplanes the fuselage was not even enclosed, but simply provided a framework of trusses and wires that held the whole thing together.

Fig. 1.3. A high-wing two-place tail-dragger, about to land. Wing struts, seen here, add structural strength to most high-wing aircraft.

Fig. 1.4. An open cockpit biplane about to land. There is a moderate headwind, nearly strong enough to make the windsock stand straight out. Airplanes land into the wind in order to reduce the speed at which they touch down onto the runway. Here the pilot's and passenger's heads are clearly visible. The propeller was stopped by the camera shutter, not by the pilot.

To get an airplane back out of the air safely, some kind of landing gear is needed (e.g., wheels, floats, skis), but only two things are really essential for flight, and all airplanes have them. These are the wings and the empennage (tail assembly). The wings are above the cabin in "high-wing" aircraft, below the cabin in "low-wing" aircraft (Fig. 1.5), and in both locations in biplanes. It makes little or no difference where they are. It is the movement of the wings through the air that creates the "lift" that makes the airplane fly. Notice how large the wings are, compared with the fuselage; a typical small airplane might be 27 feet from nose to tail, but even longer (perhaps 36 feet) from wingtip to wingtip. The wings might average about 5 feet from leading edge to trailing edge, for a total area of some 180 square feet. If this typical light airplane weighs 2300 pounds loaded, and the wings have an area of 180 square feet, each square foot of wing surface will have to support 2300/180 = 12.8 pounds of airplane. This number is called the "wing loading"; it is the amount of lift that each square foot of wing has to generate, in order to support the weight of the airplane against the force of gravity. How it does that will be explained later.

Each wing has a small movable portion on the outboard section of its trailing edge, called an "aileron" (Fig. 1.6). In flight, if the pilot deflects one aileron upward into the airstream, that wing is pushed down by the pressure against it. When one wing is low and the other high, we say the airplane is "banked", and it will

Fig. 1.5. A typical four-place low-wing airplane with tricycle gear. It is tied down securely in its assigned parking spot, a rope to each wing and one to the tail. The runway is visible just behind the airplane, and the windsock shows a fairly strong wind blowing from left to right.

turn in the direction of the low wing. Most airplanes also have inboard sections of both wings that can move out and down, extending the wing surface downward; these are called "flaps". Fig. 1.7 shows a fully extended wing flap on a low-wing airplane, Fig. 1.8 shows the same thing on a high-wing airplane. The flaps on both sides always move together. Flaps provide a means of flying more slowly by creating additional drag and at the same time the added downswept wing surface produces more lift by causing additional downwash of the air flowing past the wing. Flaps are used to assist a pilot in landing, in slow flight, and in certain types of takeoffs; but they are not used during ordinary flight, and they are not essential under any circumstances.

The tail assembly (empennage) consists of a small flat wing called a "horizontal stabilizer" and an upright vertical surface called a "vertical stabilizer" (Fig. 1.9). These tail surfaces provide a counterbalance to the wings and the weight of the engine, keeping the airplane stable and steady during flight. Sometimes the horizontal stabilizer is positioned at the top of the tail (Fig. 1.10). Like high wing vs. low wing, the exact position of the horizontal stabilizer doesn't matter a great deal. At the trailing edge of the horizontal stabilizer is a movable part called an "elevator" (seen well in

Fig. 1.6.   The ailerons. In this view of the right wing of a high-wing aircraft, the aileron (out on the outboard portion of the wing) is deflected up. Simultaneously (but not seen here), the left aileron is deflected down. In the configuration shown here, if the aircraft were flying, the right wing would be pushed down by the airstream, producing a bank and turn to the right.

Fig. 1.7. The flaps. This low-wing retractable has its flaps fully extended. Here we see the left flap, on the inboard section of the wing, with the tip of the right flap just visible underneath the airplane.

Fig. 1.8. Extended flaps on a high-wing aircraft. The left aileron can also be seen; it is the trailing outboard edge of the wing, adjacent to the flap.

Fig. 1.9), which pushes the entire tail down or up according to whether it is angled up or down into the airstream. Suppose the elevator is in the down position, as in the figure. The airstream will push against it underneath, forcing the tail upward, and thus tipping the nose downward. In some aircraft the entire horizontal stabilizer (rather than a separate elevator) tips up or down for the same purpose.

Fig. 1.9. The empennage (tail assembly). Here we see the horizontal stabilizer, fixed to the fuselage, and the movable elevator (here drooping downward) at its trailing edge. The rudder, hinged so it can move from side to side, is attached to the trailing edge of the vertical stabilizer, much like a ship's rudder.

Attached to the trailing edge of the vertical stabilizer is a movable flat surface called the "rudder" (Figs. 1.9 and 1.10). Much like a ship's rudder, it makes the airplane yaw to left or right as it is angled into the airstream. For example, if the rudder is angled to the right, the airstream forces the tail to the left. By now you have probably figured out (correctly) that the rudder, the elevator, and the ailerons all work on the same principle—using the pressure of the airstream to change the airplane's position in space. Thus, airstream against a left-deflected rudder forces the tail to the right. Airstream against an up-deflected elevator forces the tail down. Airstream against an up-deflected aileron forces that wing down.

THE ENGINE DOESN'T MAKE IT FLY—IT'S THE WINGS!

I shall never forget how I unintentionally frightened a good friend on his first flight. The first half hour had gone famously. Bob had thrilled to the experience of lifting off the runway in a small airplane—an exhilarating feeling I still get on every takeoff. It has something to do with a sensation of freedom, with the fantasy of actually having wings, of leaving behind all earthly concerns. Then we both enjoyed the lush green California hills beneath us, as they seemed to roll down to the ocean, where the surf crashed against the rocks.

Fig. 1.10.   A low-wing aircraft with horizontal stabilizer positioned high on the tail.

Cruising along at 3000 feet, I decided to go down nearer the ocean surface and fly along the beach at a few hundred feet. Without thinking, I just pulled off the power, shutting down the engine. I'll never forget the look of terror on Bob's face, and his frightened gasp. Why hadn't I warned and reassured him in advance that we could perfectly well glide down without power? It was so natural to me, I never gave it a thought. But people like Bob, who are not familiar with aviation, fear that if the engine should fail, the airplane will fall out of the sky. These people assume that the engine and propeller are essential for flight. But they are wrong. A glider flies very nicely without either (Fig. 1.11). A few thousand feet up in the air, a glider moves along smoothly at 40 or 50 k*, and it certainly doesn't fall out of the sky. What creates the upward force to counteract gravity and keep the glider from falling? And what keeps it moving along without a propeller to drive it through the air? The force that moves an aircraft forward through the air mass is called "thrust"; the propeller produces it in a powered aircraft, gravity produces it in a glider. Actually, the glider "tries" to fall, but the pressure of the air against the wings prevents that. Instead, the easiest path is a forward movement, associated with a slight fall, i.e., a certain loss of altitude with each foot of forward movement. You don't get anything for nothing. You pay for the forward movement by losing some altitude, you pay for not falling faster by moving forward. Moreover, you paid for this glider flight in the first place by the cost of the energy that towed you a few thousand feet into the air. One could say, rightly, that the tow plane is the glider's detachable engine.

Note that even though the glider descends constantly through the surrounding air, it may actually gain altitude if that air is rising. As a matter of fact, if that were not true, very few people would bother with gliding. The exciting part of gliding is finding updrafts that will lift one higher and higher. In good gliding country, where wind sweeps continuously upward at a ridge line, the glider can remain aloft—and very high aloft—for many hours. While it is gaining altitude, the glider is in descending flight with respect to the air

---

*In this book "k" is the abbreviation for "knots", nautical miles per hour. A nautical mile (abbreviated "NM") is 6,080 feet whereas an ordinary statute mile is 5,280 feet, a difference of 15%. In the present example, 50 k would be 50 × 1.15 = 58 mph.

around it. This concept may be difficult to grasp; we shall deal with it again shortly in connection with the idea of the **relative wind**.

If the air is perfectly still—without updrafts or downdrafts—how much can the glider move forward for each foot of altitude it loses? What is the **glide ratio**? For a simple training glider, it is about 20 to 1, i.e., 1 foot down for every 20 feet forward. Notice that if you start a mile high in such a glider, you can land 20 miles away, and it will take you nearly a half hour to get there. Not exactly falling out of the sky! And more advanced sailplanes have a glide ratio almost twice as great as that. The biggest glider is the space shuttle, which has a gliding range of thousands of miles from the point where it leaves orbit to reenter the atmosphere and eventually make a perfect landing without power on the designated runway.

Any airplane without its engine running becomes a glider. It's not as good as a real glider—it wasn't built to be—but its glide ratio

Fig. 1.11. A glider approaching to land. It flies nicely without an engine or prop, but only in a descending mode. The windsock shows that the surface wind is calm.

is not too bad either, about 9 to 1. Climb to a mile high and turn off the engine; you can glide about 9 miles to a landing, and it would take you about 8 minutes to reach the ground. Again, not exactly falling out of the sky!

Pilots practice such "engine-out" routines in order to develop and maintain their skills at finding good landing spots in an emergency. Ask a pilot friend to demonstrate this, repeating what I did with Bob, and exercising due caution not to fly dangerously low. See and feel for yourself how comfortably an airplane flies with its engine idling. Of course, one would never make an actual off-airport landing except in a real emergency; in the demonstration, the engine will be brought to full power, at a safe altitude above the terrain, in ample time to climb out again.

So what does the engine really do? It does not make the airplane fly, the wings do that. The engine furnishes the power needed to overcome the force of gravity. Thus, it determines the **mode of flight**— level or climbing or descending. With the engine idling, the airplane descends in a glide. With moderate power, it can fly level, maintaining a constant altitude. And with excess power over that, it can be made to climb.

The engine works by spinning the propeller. As the propeller spins, it scoops air and drives that air backward. This pushing back of enough air, fast enough, is the **thrust** that moves the airplane forward. It is very similar to the way a propeller drives a ship forward through the water. The fact that a ship's propeller is in the rear while an airplane's propeller is in front is irrelevant. The airplane propeller can just as well be mounted in the rear (reversing the angle of the blades, of course); in fact, there are some pusher-type airplanes. For that matter, a ship's propeller could just as well be mounted at the bow to pull the ship through the water. Remember, air is a fluid, like water. In both cases the blades are angled to drive the surrounding fluid backward, so that the ship or plane is forced to go forward. If you have ever seen the dust fly behind an airplane taxiing on the ground, or if you ever happened to be standing behind one as it rolled past, you will have no doubt about the propeller pushing air backward. In level or climbing flight the forward movement of the wings through the air is caused by the engine producing thrust; and it is that movement of the wings through the air that creates the force that keeps the airplane from descending. That force is called **lift**.

To understand, in a somewhat oversimplified way, how lift is produced, think of running along with a kite. The kite is set at an angle to its own forward motion, and therefore, as you pull it through the air, you create a pressure on the under surface that is greater than the pressure on the upper surface. If a wind is blowing, you can stand still and let the wind do the work. The result is the same, the kite continues to fly as long as you hold the string so that the wind can create pressure against the under surface.

Notice an important point here. What allows the wind to press against the lower surface of the kite is the fact that you are holding the string. When you were running, you created the wind for the kite by pulling. Now you hold the kite while an actual wind pushes on it. What matters in both cases is the **relative wind**, the wind relative to the kite. If the kite could "feel" the relative wind blowing on it, there would be no way for it to tell whether you were pulling it through calm air or whether you were holding it stationary in an actual wind. We shall return to the idea of relative wind in connection with airplanes. Just note here that if you let go of the string, the kite flutters and falls at once—it loses all its lift. If there is an actual wind, the kite is blown along with the wind, but it flutters and falls nonetheless, because there is no longer a relative wind to press on its lower surface and produce lift.

Now consider a balloon, not held like the kite, but drifting along. Here lift is not created by a wind, but internally by a lighter-than-air gas or by hot air (which is lighter than the surrounding cool air). No matter how strong the wind is on the ground, the people in the basket of a balloon feel no wind at all. They are carried along in the moving mass of air, which on the ground we feel as a wind; but the balloon and its occupants experience no relative wind. A flag in the gondola would hang limp. The balloon has no motion relative to the mass of air in which it is immersed, for it has no propeller to drive it through that air mass.

Consider one more example—one you can almost "feel". Suppose you are wading, waist-deep, in a fast-moving stream. You can feel the pressure of the water rushing by, but the instant you lift your feet, you become part of the flow, and you feel no pressure at all. The stream is still flowing as it was before, but you have become a "balloon"—the "relative wind" pressing on you has dropped to zero.

Let's return now to the airplane. The engine spins the prop and thrusts the wings forward through the air, just as you did running with the kite; and the forward motion relative to the air creates a relative wind. The wings are set quite precisely at a slight angle (as was the kite) so that the relative wind produces more pressure on the bottom surfaces than on the top. This pressure creates lift, so that the airplane doesn't fall, but can stay level as it moves forward. As I pointed out in an earlier section, the relative wind has to be strong enough, with a typical general aviation airplane, so that each square foot of wing surface lifts around 13 pounds. The engine must produce enough power to move the airplane forward fast enough to generate enough relative wind to produce enough lift to support the weight of the airplane. Any leftover power can be converted to speed or to climb capability. There you have the subject of basic aeronautical design in a nutshell. The performance of an airplane depends upon the engine power (for level flight or climbing), the propeller design, the wing surface area, the angle at which the wings are set, and the air resistance (drag) that has to be overcome.

INSIDE THE CABIN: THE POWER CONTROLS AND THE POWER INSTRUMENTS

Now it's time to climb into the cabin. Sit in the right front seat of an airplane. If your pilot friend flies a complex airplane, it may be better to make your first acquaintance with the controls and instruments of a simple trainer. That should be easy to arrange.

In front of you and to your left is an array of dials, knobs, levers, and switches—perhaps a bit bewildering at first. Let's start to sort these out. Your pilot friend can help you. Begin by identifying the power controls and engine instruments. First find the **throttle**, a push-pull knob (in some airplanes a lever) that moves forward and back (Figs. 1.12 and 1.13). If it's a twin, there will be two throttles—one for each engine (Fig. 14.). The throttle controls how much air-and-fuel mixture passes into the cylinders to be burned, very much as the gas pedal does on a car. Throttle pulled back (like foot pulled up/back from the car accelerator pedal) is closed throttle; the fuel-air mixture to the engine is almost shut off, and the engine idles at a very low speed. Throttle pushed forward, open throttle (stepping on the gas) makes the engine run faster.

Next to the throttle is another push-pull knob or lever, which is always colored red, called a **mixture control**. Like the "choke" on old cars, it controls how rich a fuel mixture flows into the cylinders. Pulled all the way back (full lean position) it shuts off the fuel flow entirely, so that the throttle can only allow air to enter the cylinders. Pushed all the way forward (full rich position), it provides a maximum flow of fuel. The way to shut down an engine competely is to pull the mixture control all the way back, and that is exactly what the pilot does after parking the airplane at the end of a flight. The other use of the mixture control is to produce a leaner mixture at the higher altitudes. The reason for this is that the higher you go, the less dense is the air, so that less oxygen flows into the cylinders on every stroke of the piston. You can't help that, so to keep the right proportion of air to fuel for most efficient burning, it is necessary to make a proportionate reduction in the fuel flow.

Fig. 1.12.   The power controls in a simple single-engine airplane with fixed-pitch prop. Three control knobs are seen—the round black throttle knob in the middle (marked THROT PUSH OPEN), the knurled bright red mixture knob on the right (marked MIX PULL LEAN), and the square carburetor heat knob on the left (marked CARB HT).

This is a point worth noting. If you were to begin flying an airplane from the right seat at some thousands of feet above sea level, you would find that the pilot had the mixture control part-way back. When you descend, the mixture will have to be enriched (pushed forward), bit by bit, until at sea level it is full rich (all the way forward).

The third control shown in Fig. 1.12 is the **carburetor heat**. Airplanes with carburetors may sometimes develop an icing condition in the narrow carburetor throat. This results in decreased power, which may make the engine sound uneven ("rough"), and which also results in decreased power indications on the RPM or MP gauges (see below). Pulling out the carburetor heat knob melts the ice and restores power. Carburetor icing is prone to occur on moist days and at low power settings, so it is customary to apply carburetor heat whenever low power is being used, as in the approach to

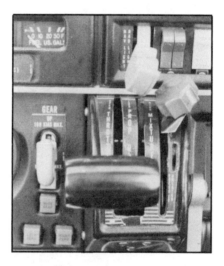

Fig. 1.13. The power controls ("power quadrant") in a single-engine airplane with variable-pitch (constant-speed) propeller. In this arrangement the throttle is the black lever with the large handle at the left. The prop control is the gray knob between throttle and mixture control. The mixture control is knurled and red, as before. Airplanes with fuel injection engines have no carburetor, so—as here—there is no carburetor heat control. As this is a retractable-gear aircraft, there is a gear handle, seen to the left of the throttle, with up and down positions clearly marked. The three square gear indicator lights just below the handle will show all green when all three landing gear are down and safely locked in position for landing.

landing. Some airplanes, like some cars, don't have a carburetor, but use fuel injection instead. If you can't find a carburetor heat control, that is probably the reason.

Between the throttle and mixture controls there may be a third knob or lever—a **prop control** (Figs. 1.13 and 1.14). You will recall from our walk around the airport that there are two kinds of propeller. Only airplanes with variable-pitch (constant-speed) props will have a prop control. The less expensive airplanes have the simpler fixed-pitch system, in which more power means higher propeller speed (RPM). With a variable-pitch prop, the pilot can set a certain propeller speed. Then the pitch of the prop blades automatically adjusts, to maintain that RPM. This allows for more efficient and quieter flight. The blades can be at a flatter angle for high-power high-RPM use, as in takeoffs; or they can turn over more slowly but take deeper bites into the air when lower power suffices. Right now, from your standpoint as a right-seat occupant, you can forget about the prop control. In most aircraft you can just push it all the way forward and leave it there; the airplane will fly fine at all power (throttle) settings, and the small sacrifice of efficiency will be of no importance. You can check to be sure if this is safe by noting the RPM reading on the tachometer, as described below.

Fig. 1.14. The power controls in a twin with variable-pitch propellers. The arrangement is exactly like that in Fig. 1.13, but now there are two of everything, one for each engine.

The **tachometer** (Fig. 1.15) shows, in hundreds of revolutions per minute (RPM) how fast the prop is spinning. A normal reading in flight might be 24, meaning 2400 RPM, in the green shaded band indicating normal cruising values. A helpful standard convention in the marking of dials and gauges is that normal ranges of values are shown by green bands or arcs. If you advance the prop control full forward, as suggested above, check to make sure the RPM indication is not higher than the top of the green band. With the engine idling the reading might be around 8, meaning 800 RPM. The propeller is attached directly to the engine driveshaft, without gears, transmission, or clutch.

There is a meter in association with the tachometer (at bottom in Fig. 1.15), which registers how many hours the engine has been running at normal speed since it was new. At a normal speed of say, 2500 RPM, the prop will turn 150,000 times an hour. To avoid huge numbers, the "tach time" will read in hours and tenths of an hour rather than in total revolutions; in this example the reading would increase by one-tenth of an hour every 15,000 revolutions. At a slower speed, it will increase proportionately less, at a faster speed proportionately more. In some airplanes there is a meter ("Hobbs meter") that tells the actual clock time the engine has run, in tenths

Fig. 1.15. The tachometer, showing hundreds of revolutions of the propeller per minute. In typical flight it would read between 21 and 26 (i.e., 2100-2600), and this normal range is marked by a green band. The hour meter, showing cumulative time (in hours and tenths of hours), is also seen.

of an hour, regardless of RPM. Airplane engines are overhauled at regular intervals, according to the tach or Hobbs meter time.

An airplane can not have a true odometer showing distance travelled. When the airplane is flying, it is disconnected from the earth over which it passes. It moves through an ocean of air, just as a ship moves through an ocean of water. All we can measure is the speed of the vehicle (airplane or ship) through the surrounding fluid (air or water). If a ship is travelling along with a current, it covers more distance over the ocean bottom in a given time than if it is travelling against a current. In the same way, if an airplane is travelling along with a current of air—which you would call a wind if you were on the ground—it covers more distance over the earth's surface in a given time than if it is travelling opposite to the movement of the air mass.

Imagine that an airplane capable of making 100 k is flying along in the same direction as a 100-k wind—a very strong wind but not unheard of at high altitudes. The airplane moves at 100 k **within** (relative to) the air mass, and it is also carried along at an additional 100 k **by** the air mass. Thus, it covers 200 NM over the ground each hour. Now imagine a balloon in the same air mass. The balloon has no motion whatsoever within the air mass, but it is carried right along with the wind, so it travels 100 NM each hour. Finally, let's have an airplane fly directly into the wind. It still moves 100 k within the moving air mass, but now it covers no distance at all over the ground; after an hour it could land in exactly the same place where it started. That can actually happen. I once took off from Boulder, Colorado, intending to climb up over the Continental Divide and head home for California. The higher I climbed, the less progress I seemed to make, as the headwind became stronger. Finally, after about half an hour, I looked straight down and there was the same airport I had started from. I decided it would be wise to land. The next day the strong wind had subsided and I flew home uneventfully.

The experience described above raises an interesting point about flying. The air is invisible, and its movement—the wind—is also invisible. In addition to the steady movement of large air masses, which we call wind, there are local air currents moving up, down, and sideways. If you fly into an upward moving column of air, you will be suddenly bumped, as your aircraft is lifted. A downward moving column will also give you a jolt, and you will seem to drop—in popular jargon an "air pocket". Since the air is rarely perfectly calm, one has to learn to enjoy these mild up and down movements, much as you would in a sailboat. There is almost always good advance warning about the kinds of weather that produce really uncomfortable turbulence, and then we try to avoid flying altogether.

How can you tell how much power the engine is developing? First look at the power controls. If you can only find a throttle and a mixture control, the airplane must have a fixed-pitch prop, and then the tachometer (RPM gauge) will tell the whole story—the higher the RPM, the greater the power. On the other hand, if there is a prop control, it will set the RPM to a fixed value, so the tachometer will tell nothing at all about the power being delivered by the engine. Instead, there will be another instrument, called a **Manifold Pressure** (MP) gauge (Fig. 1.16), which will show engine power directly. Once the engine is running, the bigger the number, the higher

Fig. 1.16. The manifold pressure (MP) gauge. This engine is not running. When the engine is running, higher numbers mean higher power. Then, if the throttle is pulled back to idle, the MP will drop to around 12; at full power it may read around 25-30.

the power. If the engine is not running at all, however, the MP gauge simply shows the atmospheric pressure in inches of mercury, typically about 30 at sea level, as in Fig. 1.16.

THE FLIGHT CONTROLS AND THE FLIGHT INSTRUMENTS

The principal flight controls are combined into a single unit, the **yoke**, which looks something like a steering wheel (Fig. 1.17). The dual controls operate together so the airplane can be flown equally well from either front seat. Some airplanes (especially older models) have a "stick" mounted in the floor instead of a yoke for performing the same functions. Turning the yoke (or moving the stick) to the left initiates a left turn. to the right a right turn. Remember the ailerons, which force one wing down? Turning the yoke a very small amount in one direction causes the aileron on that side to flex upward and at the same time the opposite aileron to flex downward. The result is that the airplane banks in the direction of the up aileron (which is the direction the yoke was turned); that wing

Fig. 1.17. The yoke. Turned this way, the airplane would bank very steeply to the right. Tiny yoke movements suffice. At the left edge of the picture the pilot's yoke is just visible.

drops (forced down by the airstream), and the other wing rises (forced up by the airstream).

Once in a bank, and turning, you have to neutralize the ailerons (i.e., return the yoke to its original centered position) in order to continue turning with the same angle of bank. This is quite different from what happens in a car or a bicycle, where you turn the steering wheel or handle bars to turn the vehicle, and straighten them out to stop the turn. The airplane doesn't "know" whether it's flying in a banked circle or whether it is flying straight and level. With ailerons neutral, it will tend to continue doing whatever it is doing, and thus to remain in a banked turn until you do something about it.

To stop a turn, therefore, you have to unbank, i.e., roll back to a wings-level attitude, by briefly applying opposite aileron (i.e., opposite yoke movement to what you used for rolling into the bank). This lowers the high wing and raises the low wing. When the wings are level again, you return the ailerons to a neutral position once more. Thus, turning an airplane to head in a new direction involves four small actions in sequence—rolling into the bank, neutralizing ailerons, waiting for the desired amount of turn to be completed, then rolling back to wings-level attitude, and finally neutralizing ailerons again. Student pilots practice all this in steep banks and steep turns to learn good coordination, but there's no need for you to do that. In fact, now that you know about the four sequential actions, you can practically forget about them. If you make **very small** yoke movements **very delicately**, your banks will be very slight, and your turns will be gentle and slow. Just use tiny yoke movements to do whatever is needed to initiate turns, and whatever is needed to stop them when you've turned enough.

There is a simple secret to flying smoothly, and this is a good time to tell you about it. The airplane is a very stable machine, it tends to continue doing whatever it is doing. Flying level, it tends to continue flying level. In a banked turn, as we have seen, it tends to continue turning. The secret is always to handle the yoke gently, with an extremely soft touch. I teach my students to use only a thumb and forefinger on the yoke, never to grip it tightly. This takes advantage of the airplane's natural stability by letting it pretty much fly itself. Students are convinced that their instructor is some kind of super-pilot, because as soon as the instructor takes the controls, the airplane settles down to perfectly stable flight. I don't want to dispel the pleasant illusion, but the trouble was simply that the student was fighting with the airplane, pushing and pulling and turning and twisting, while the instructor knows enough to just leave it alone. In a nutshell, you don't **fly** the airplane, you **guide** it.

Another secret for the right-seat flyer is that the rudder pedals can be ignored completely, except on the ground, where they are needed for steering and braking. We'll discuss that later. Your pilot friends may be surprised at this advice to ignore the rudder pedals, because they have been subjected to a lot of training in the proper use of rudder in various phases of flight. From your point of view right now, however, the rudder is best forgotten. It serves no absolutely essential purpose, it is only a refinement that takes the rough edges off some of the airplane's behavior. Keep your feet on the floor.

This is a good time to locate the instruments that will tell you if the airplane is flying straight, with wings level, or if it is in a banked turn. There are three useful ones. First, there is an **attitude indicator** (AI, also called "artificial horizon", Fig. 1.18). In this clever gyroscopically driven device you see a little mock airplane (representing your airplane) positioned against a symbolic horizon. If your airplane banks to the right or left, the mock airplane seems to bank correspondingly. If your airplane is in a nose-high attitude, as when climbing steeply, the mock airplane is positioned well above the horizon line. Actually, the instrument is constructed so that the mock airplane is immovably attached to the instrument panel, while the horizon line behind it does the moving. This will not cause any confusion if you just look at the mock airplane and see what it does **against the background**. To aid that perception, most attitude indicators have a light blue area above the horizon bar (representing sky) and a dark area below it (representing the ground). For example, if you see the mock airplane with its left wing against the blue sky and its right wing against the dark ground, you know you are banked to the right. So even without looking outside, you can tell from the AI what your airplane is doing. You can even use the AI to start turns, stop turns, level the wings—all without looking outside. The AI is the most important of the indicators for instrument flying (called IFR for "instrument flight rules").

The second instrument that tells you if you are flying straight ahead or turning is the **turn needle** (Fig. 1.19) or **turn coordinator** (Fig. 1.20). These devices are also gyroscopic. Unlike the AI, they don't necessarily tell anything about whether your wings are level or banked; they only sense if you are flying straight or turning. Usually, however, it amounts to the same thing, since turning is normally accomplished by banking. In each instrument there is also a free-floating ball in a tube of fluid, which remains centered in correct, well-balanced, flight. Like the rudders, you can forget about the ball for the present, it is of no importance to you.

The turn needle shows a left turn by moving to the left—the faster the turn (usually meaning the steeper the bank), the greater the deviation from the central upright position. The turn coordinator shows the same thing but displays the turn by the symbolic airplane tipping in the direction of the turn. If you watch both the turn coordinator and the artificial horizon while taxiing, you will see at once how they differ. When the airplane turns on the ground, since it doesn't bank, the artificial horizon remains level; but the turn coordinator (or turn needle) shows the turn.

Fig. 1.18. The attitude indicator (AI). At left, the little mock airplane is banking and turning to the right, and it is also in a very nose-high attitude, as in a steep climb. The airplane at right is in a left turn; it is in level flight, so the mock airplane is right on the artificial horizon (which divides the dark simulated ground from the light simulated sky). The wedge-shaped pointer at top shows how much the airplane is banking, at left about 10 degrees, at right about 20 degrees.

The third instrument that reacts in turns is the **heading indicator** (HI, Fig. 1.21). It is a gyroscopic compass, marked from N (north, 0 or 360 degrees) around through E (east, 90 degrees), S (south, 180 degrees), W (west, 270 degrees), and back to north. Notice that a zero has been dropped from each number; thus, for example, 30 means 300 degrees. If your heading is north and you start a left turn, the HI will show you passing through 35, 34, 33, etc. as you turn toward the west. In a right turn, toward the east, you will pass through 1, 2, 3, etc. The airplane in Fig. 1.21 is flying on a heading of 250 degrees, just a little south of west. The symbolic airplane on the face of the dial represents your airplane. To turn this airplane to north, for example, will require a gentle right bank, held for long enough to accomplish 110 degrees of turn—20 degrees to west, then 90 degrees more to north.

Finally, there is an old-fashioned **magnetic compass** (Fig. 1.22), always mounted somewhere up near the windscreen. Under ordinary circumstances this is not used, except for one purpose. It can be relied upon in straight and level flight, and therefore it can be used to set the gyroscopic HI correctly. The HI tends to drift, and therefore requires periodic re-setting, sometimes as often as every 15 minutes. Except for this slow drift (called "gyroscopic precession"), the HI is very reliable under all flight conditions. The magnetic compass, on the other hand, is correct **only** when the airplane is flying straight and level (i.e., not turning, climbing, or descending), at a constant airspeed. It gives erroneous readings whenever the airplane banks, climbs, descends, or even just speeds up or slows down, so one never relies on it directly when a working HI is at hand. However, some small aircraft do not have gyroscopic instruments; then the pilot has to use the magnetic compass for heading information.

You may be wondering why these instruments are needed at all. Can't the pilot just look outside and fly by the natural horizon and by reference to the ground? Why would they need an instrument to tell them if the wings are level? There are two reasons for being familiar with these instruments. First, on a dark night they can be essential, because unless there are lighted towns and highways below, it is easy to mistake stars for lights on the ground, and thus to become confused about up and down. The same thing can happen

Fig. 1.19. The "needle and ball". The turn needle points upright when the wings are level and the airplane is flying straight. It points left in a left turn, right in a right turn. For properly slow and gentle turns, the turn needle should never be more than about half a needle's width from the center.

Fig. 1.20. The turn coordinator. Here the mock airplane banks when the real airplane turns. When the wingtips point to the "L" or "R" marks, the airplane is turning at the rate of a full circle in 2 minutes to the left or right, respectively. In a properly slow and gentle turn, the symbolic wing tip of the turn coordinator should never go beyond the "L" or "R" mark.

Fig. 1.21. The heading indicator (HI). The indicated heading, under the arrow at the nose of the symbolic airplane, should correspond to what the magnetic compass says during wings-level stable flight. If it does not, the HI can be adjusted by pushing and turning the little adjustment knob at lower left.

if a flight gets into an area of poor visibility because of deteriorating weather. Secondly, you may some day be on an actual IFR flight with an instrument-rated pilot, flying in the clouds, where these instruments are absolutely essential.

I remember a flight, long ago, when I was in training for my own instrument rating. I was flying "under the hood", a device that shuts out everything but the instruments from the field of view. My instructor was in the right seat. We were making an instrument landing approach at Fresno, California, on one of those days when smog and fog combine to blot out everything. I was unaware of anything outside, and the instruments told me everything was going fine, we were lined up with the runway, and we would be landing in a few minutes. Suddenly, my instructor became very agitated, and made as if to grab the yoke, shouting at me as he did so. Then just as suddenly, he relaxed again and explained. He had suddenly experienced "vertigo", a kind of dizziness where you have an overpowering sensation of turning upside down or diving. A glance at the instru-

Fig. 1.22. The magnetic compass. As explained in the text, this instrument is more confusing than helpful except in perfectly stable level flight. When the airplane turns, the compass even seems to turn confusingly in the wrong direction. The one shown here indicates a heading of about 070 degrees. If the airplane turns to the right (for example 20 degrees to a heading of east), the compass will seem to turn to the left. The cards full of compass corrections should be ignored completely; they are useless, as no possible correction is big enough to matter to you.

ments straightened him out at once. The point of this story is that those instruments are to be trusted and depended upon—always. They are right, and anything you feel is wrong.

The problem is that we have no way to feel our position in space other than by feeling the force of gravity. And in an airplane, gravitational force usually acts straight downward, pressing you into your seat, even when the airplane is turning, climbing, or diving. And the semicircular canals in our inner ears were not meant for flying either, they give us correct information about our equilibrium in ordinary life activities, but definitely not in flying. The first task in training people for IFR flying is to teach them to depend **absolutely** on the instruments, regardless of what they "feel".

Now let's return to the yoke and see what it does in flight. Besides controlling turns, the yoke controls the airplane's **attitude**. By "attitude" is meant the pitch, i.e., whether the nose is high or low relative to the tail, whether the airplane is pointing up at the sky or straight at the horizon or down at the ground. Usually we want the airplane to fly level. Looking out and around, we should see the airplane seeming to head for the distant horizon. Looking at the AI, we should see the little mock airplane level or nearly level on the artificial horizon.

Pulling or pushing (very gently, very slightly) on the yoke will change the pitch attitude. Pulling back toward you flexes the elevators (or the whole horizontal stabilizer) upward into the relative wind, pushing the tail down and thus establishing a nose-high attitude. Pushing forward does the opposite, and the nose goes down. For any power setting (including idle power), the more nose-up the pitch, the slower the airplane flies; and the more nose-down the pitch, the faster. This has to do with the angle of the wings to the relative wind. The more nose-up the attitude, the greater the angle at which the wings meet the relative wind, the greater the drag, and the lower the airspeed for a given power setting.

There is an important little device that controls the forward or backward action of the yoke when you are not touching it at all. This device is called the **pitch trim**. It can be a notched wheel somewhere on the floor, or in the center of the panel (as in

Fig. 1.23), or it can be a crank handle in the cabin ceiling or wall. It provides a simple, safe, and delicate way of controlling the airplane's attitude. Trimming for a more nose-up or nose-down pitch has exactly the same effect as pulling back or pushing forward on the yoke. **It sets the airspeed.** You can even think of the trim wheel or crank as having airspeed numbers displayed, such as 80, 90, 100, 110 k, a certain speed corresponding to each position.

Again the right-seat flyer can do something more simply than the pilot ever learned to. Pilots are taught to use the trim wheel only after they have first made a change in pitch by pulling or pushing on the yoke. That makes sense for pilots, because it trains them to acquire a better "feel" for the airplane. But it certainly doesn't make sense for you. Use the trim wheel (or crank) to set whatever airspeed you like, and never pull or push on the yoke at all. Try to keep your hands off the yoke entirely, except perhaps for a thumb and forefinger, and only use those to keep the wings level or to start and stop turns. This will be a safe procedure under all flight conditions. It won't do, however, for landing the airplane, when you will have no choice but to handle the yoke. That technique will be described at length in Chapter 5.

The instrument that monitors airspeed is the **airspeed indicator** (ASI, Fig. 1.24). Modern ones read in knots, older ones (as in Fig. 1.24) in statute miles per hour. The ASI has a green arc, which shows the range of safe airspeeds for normal flight. The airspeed is the rate at which the airplane is travelling through the air mass, or the rate at which the relative wind is moving past the airplane (which is the same thing, as explained earlier). You will recall that the wings can only generate enough lift if they move through the air fast enough, i.e., if the relative wind is strong enough. This is why it is very important to keep the airspeed comfortably (at least 15 k) above the bottom of the green arc on the ASI. In Fig. 1.24, for example, the bottom of the green arc is at 80 MPH, so the airspeed should not be allowed to fall below about 98 MPH. You can manage that with the trim wheel and an appropriate power setting, as described below.

The ASI is a remarkably simple device. Outside the airplane and pointing ahead is the "pitot tube"; ask your pilot friend to show it

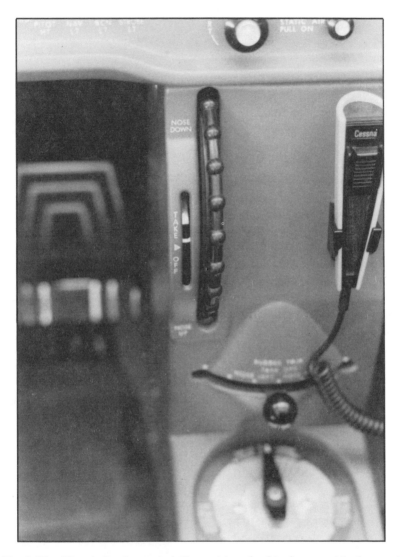

Fig. 1.23. The pitch trim control. Here a trim wheel is shown, set in the panel between the two front seats. In this view, the pilot's right-foot rudder pedal can be seen on the floor at left; a hand-held microphone is at right. Turning the wheel up (so the little ridges move up and away from you into the panel) makes the nose go down and the airplane fly faster. Moving it down and toward you raises the nose and the airplane flies more slowly. Just to the left of the trim wheel is a trim indicator, which shows if the trim is neutral (in the takeoff range) or nose down, or nose up. Another type of pitch trim control is a crank located overhead.

to you. The impact of air striking the opening of this tube presses on a diaphragm inside the airplane, which is connected to the ASI needle. The faster the airplane, the more airstream impact. Some kinds of ASI, like the one pictured in Fig. 1.24, have a movable rim adjustment that permits one to read the "true airspeed" in addition to the "indicated airspeed". Forget it! **Indicated airspeed**, shown on the fixed face of the dial, is what matters to you; it is the airspeed "felt by" the wings in generating lift, which is the same as the impact of the airstream on the pitot tube. True airspeed differs from indicated airspeed as altitude and temperature vary. If you're interested, ask a pilot friend to explain it further. Note that no simple instrument intrinsic to the airplane itself can measure the groundspeed, for the same reason that the airplane can not have a mileage odometer. However, there is a special radio device called "distance-measuring equipment" (DME), which is used in conjunction with a ground station, and which will be described in Chapter 2. There is also an extremely sophisticated and expensive device called "inertial navigation system", used chiefly by the airlines, which will not be described in this book.

Fig. 1.24.   The airspeed indicator (ASI). The outer adjustable portion of this instrument is of no concern to you, ignore it. Read the numbers in the fixed portion of the instrument face. The instrument shown here reads in statute miles per hour and also (in the little window) in knots. This airplane is flying at 160 MPH, about 140 k.

A FEW SIMPLE EXERCISES—AND TWO NEW INSTRUMENTS

Get your pilot friend to take you up to altitude in an uncongested area. Have them establish the airplane in straight and level flight at a certain airspeed and with the appropriate power setting to maintain level flight at that airspeed. Now experiment. See if you can turn toward a definite direction, perhaps toward a landmark in the far distance, and then try to stop the turn accurately. Twist the yoke very slightly toward the desired side to initiate the turn. When the wing on that side goes down a bit, return the yoke to its neutral position. As the turn continues, look outside but also glance at the three useful turn instruments—the little mock airplane on the AI, the turn needle (or turn coordinator), and the HI showing the changing compass headings. Now stop the turn by briefly twisting the yoke in the opposite direction, and neutralize it again as the wings return to level. Try the same exercise to the left and to the right. Make it smooth. Try planning a certain amount of turn, or a turn to a certain heading, and practice until you can roll out smoothly where you want to.

Next make a mental note of the power setting—on the MP gauge if there is one, otherwise note the RPM on the tachometer. Reduce the power slightly by pulling back a little—just a few millimeters— on the throttle. Remember to touch the yoke only lightly to level the wings if one wing or the other shows a tendency to dip (the airplane may not be in perfect side-to-side balance). You'll recall that the function of the engine is to control climb and descent. Here, when you reduce the power slightly, the nose will drop a little, and you will begin to lose altitude, ever so slowly.

The instrument that tells you your altitude (above sea level) is the **altimeter** (Fig. 1.25, left). It reads something like a clock, with the large hand sweeping through the hundreds of feet (one full revolution represents 1000 feet), and the small hand pointing to the thousands of feet. As you start down, the large hand will begin, very slowly, to unwind. The altimeter senses the altitude above sea level by the atmospheric pressure, which is lower the higher you go. At sea level, the altimeter should always read zero. However, the actual atmospheric pressure at sea level changes with changing weather, and this must be corrected for. The little window between the "2" and the "3" shows numbers that can be set by the knob at the lower left. These numbers (called "altimeter settings") correspond to the actual sea level atmospheric pressure in inches of mercury. The one in the figure is set to 30.14. Who told the pilot to set it that way? A controller did, by radio—but more about that later. The point here is that if you set that number in the window, the altimeter will show your correct altitude above sea level.

Another useful instrument is the **vertical speed indicator** (VSI, Fig. 1.25, right). This shows the rate at which you are going down or up (or neither, in which case it points to zero) in hundreds of feet per minute. It is very sensitive to an incipient change in altitude. Thus, the moment you reduced power, the VSI began showing a descent, even before the altimeter indicated a change of altitude. In the example shown here, we are looking at the altimeter and the VSI of my airplane, flying over the Santa Cruz mountains in northern California. Let's see what we can learn from the two instruments. First, the altimeter. At a glance, it says about 300 feet (large hand) past 7000 feet (small hand). Confirming this, we see that the odd-shaped marker, which indicates tens of thousands of feet, is pointing a bit below 10,000. Pains are taken, in designing altimeters, to prevent misreading by 10,000-foot amounts; imagine

Fig. 1.25. Left: The altimeter. Here the "altimeter setting", in the little window between "2" and "3", is 30.14. Instructions for the setting are given by the controller over the radiotelephone. This airplane is at 7340 feet above sea level. See further explanation in text. Right: The vertical speed indicator (VSI). The same airplane is descending 300 feet per minute. You should never carry out climbs or descents exceeding a few hundred feet per minute.

how serious it would be if—at night or in the clouds—a pilot meant to be at a safe 17,300 feet and flew instead at 7300 feet. The altimeter can be read accurately to 10 feet, so here I was actually descending through 7340 feet at the moment the picture was taken. I said "descending". Did you notice that on the VSI? The descent is at a comfortably slow rate, 300 feet per minute; the VSI markings are in units of 100 feet per minute.

Push the throttle back in to the same power setting as before. Notice how the VSI returns to zero and the altimeter stops unwinding. Now add a little power and note how a climb is indicated on the VSI, followed by the altimeter starting to show the climb, too. All climbs and descents should be very very slow, never more than a few hundred feet per minute. Through this whole exercise the ASI will have shown little change in airspeed.

Here is a very important safety principle: In flying, **only change one thing at a time**. If you are turning, do not simultaneously make power changes or airspeed changes. If you are changing altitude, do not simultaneously change the airspeed.

Now you are ready to try some deliberate airspeed changes. Make a mental note of your airspeed on the ASI. Remember, the airspeed control is the pitch trim (wheel or trim crank). Begin with a small reduction in airspeed by trimming to a slightly more nose-up attitude. Until you are familiar with things, it is easy to get mixed up about which way to turn the trim wheel or crank. Rotating the wheel down and toward you raises the nose, and turning the crank clockwise does the same. But double-check with your pilot friend before you do anything. If possible, practice changing the trim while the airplane is tied down, and as you move the trim wheel or crank, say out loud "Nose up" or "Nose down", to fix it in your memory.

In the air, make only a small change—the trim is very sensitive. As you trim for a more nose-up attitude and a slower airspeed, something interesting will happen—the airplane will begin to climb. This is because the power setting, which was just sufficient to maintain altitude at a certain airspeed, is now excessive for the lower airspeed. Put simply, the power was divided between two functions—keeping the airplane from descending, and driving it through the air mass at a certain airspeed. With the same power and a lower airspeed, there is now an excess of power, which is converted to climbing performance. But no matter! You know how to stop the climb; just make a small power reduction. You can bring the throttle back just enough to return the VSI to zero. The ASI should now show that your airspeed is 10 or 20 k slower than before, and you should still be flying level. If you gained some altitude in that first unexpected little zoom when you changed the trim, make a slight power reduction, get back to the old altitude, then push the throttle in again as required for level flight. These throttle changes are very small—no more than a few millimeters in or out—and they should be guided by what the power instrument (RPM or MP) tells you.

Now do the opposite exercise. Establish a slightly more nose-down trim than you had originally, and you can get an airspeed 10-20 k faster than before. Of course, you will begin to lose altitude, but a little more power will stop that promptly, and level you off again.

The airplane's performance is exactly predictable. For a certain airspeed (determined by the trim), it will require a certain power setting (RPM or MP) for level flight. After a climb or descent, if you use the throttle to reestablish that same power setting, the airplane is bound to fly level again. The experienced pilot knows this and "flies by the numbers". Suppose level flight at 100 k in your airplane requires 2300 RPM, and you are flying along that way at 5000 feet. How do you climb to 5500 feet? Just push in the throttle for full power. Then, when the altimeter shows 5500, bring the throttle smoothly back to 2300 RPM again.

Try a descent now, from 5500 down to 4500 feet. Gradually reduce the power to learn what RPM (or MP) gives you a rate of descent of not more than a few hundred feet per minute on the VSI. Suppose that setting is 2100 RPM. From now on, you can always initiate a descent by reducing power to 2100 RPM, and you can stop the descent by returning the power to 2300 RPM—all the while being trimmed for 100 k.

Everything I have said so far about level flight, climbs, and descents is correct in principle, but it may not work out exactly so in practice. If the air is perfectly still, what I said will be literally true. But suppose there are some updrafts or downdrafts or both in alternation (which is usually the case because if air flows up in one place, it has to flow down someplace else). Then the airplane is bound to move with the air mass—whether up or down—and this movement will be added to (or subtracted from) what you are trying to accomplish. The best way to proceed is to have patience. If you reduce power in order to descend, a simultaneous updraft may delay your descent, but eventually you will start down. In some locales, however, "eventually" could be a long time. A continuous wind blowing upslope at a mountain range will be deflected upward, and if you are flying parallel to the ridge line, even well above it, you will have to reduce power considerably just to stay at the same altitude. In fact, if you were in a glider, you would be thrilled—this is the "ridge soaring" that glider pilots love.

These simple exercises are meant to build your confidence, and they are really all you need to know to fly an airplane safely and fairly competently—if not very elegantly. Elegance comes with a pilot's license and a lot of experience. What I have done here is give you the basic knowledge and the tools to understand how the airplane flies and how to fly it yourself from the right seat. If your pilot friend doesn't become impatient, you ought to be able—after a few hours of practice—to develop fairly good technique and to feel comfortable and relaxed about the way the airplane responds to your actions. Landing the airplane is a rather different and special business, which we shall go into fully in Chapter 5.

It remains now to discuss how to steer the airplane on the ground. The rudder pedals, which I dismissed as being of no importance to you in the air, are extremely important on the ground. They are linked to the nosewheel or tailwheel (as the case may be). Left pedal makes the airplane turn to the left, right pedal to the right. This seems backward at first to many people because if you want to turn a bicycle to the left you advance your **right** arm, not your left arm. The way you turn an airplane is different, and you just have to learn it by practice. Get your pilot friend to let you steer the airplane during taxiing, when the taxiway ahead is very wide and perfectly clear, so you can get the feel of how the pedals work. Don't be embarrassed if at first you zigzag back and forth on the taxiway, it takes a while to get the hang of it. Don't forget, your pilot friend once did it that way, too!

The brakes are operated by the same pedals—either toe brakes or heel brakes. Each wheel brake is operated independently by the corresponding pedal. Thus, to stop, you apply both brakes together. To make a sharp turn, you apply one brake selectively, so the airplane can pivot around that wheel. For example, to turn sharply to the left, you apply the left brake at the same time that you push on the left pedal. Some practice in steering on the ground and in operating the brakes is essential to prepare for the remote possibility that you may some day have to land the airplane by yourself, as described in Chapter 5.

SUMMARY

In this chapter you learned why an airplane flies, and the basic principles of guiding its flight. You also discovered what can be learned from the power instruments and the flight instruments. If you can find the opportunity to apply and practice the lessons of this chapter, you should soon feel confident about handling the airplane—flying straight and level, turning to new headings, climbing, descending, and taxiing.

Chapter 1: Problems

(See Appendix for answers.)

1. You learned how the rudder, the elevators, and the ailerons work on the same principle—by moving so the airstream presses against them.

   a. Imagine an airplane without ailerons. Could it be turned? How?

   b. Imagine an airplane without an elevator. Could it be made to climb or descend? How?

2. You learned how a glider is towed into the air and released at a couple of thousand feet. Gliders can also be launched by a catapult mechanism—a winch that quickly pulls in a cable that is attached to a hook under the glider's nose. When the cable is dropped, the glider is only a few hundred feet in the air. How does it gain more altitude?

3. You are flying on a straight course eastward. A brisk wind of about 40 k is blowing from the north, but your pilot doesn't know about it. Your destination is exactly 100 NM away and you are flying at exactly 100 k. One hour later you look around but the destination airport is nowhere in sight. Where is it?

4. How could you have helped the pilot to avoid the situation just described?

5. You have been flying along at 11,500 feet. You come to a layer of clouds. To go over the top of them and stay well clear, you would have to climber higher than 15,000 feet. Are there any reasons this might not be practicable?

6. If you turn the yoke to put the airplane in a steep bank (a hypothetical example—don't do it!), the nose will drop. Why?

7. Sit in the right seat while the airplane is tied down, with the electrical master switch turned off, and see if you can identify the following instruments and answer the questions. For some of the questions you may need additional information from your pilot friend.

    a. **The AI.** What would your airplane be doing if the AI looked like this in flight? Why does it look like this now?

    b. **The ASI.** What airspeed does it show? Is it actually zero? What would the airplane be doing if it looked like this in flight?

    c. **The magnetic compass.** What direction does it say you are heading at your tie-down? Is it correct?

    d. **The HI.** What direction does it say you are heading? Does it agree with the magnetic compass? If not, why not? If so, why? What drives the gyros that make this gyroscopic instrument work?

    e. **The turn indicator or turn coordinator.** Is it operated electrically or by vacuum from an engine-driven vacuum pump? Is it operating now? If it were in flight, what would the airplane be doing?

f. **The altimeter.** What altitude above sea level does it show? What is the actual altitude of your airport? Is the altimeter correct? If not, why not? Using the adjusting knob, make the altimeter needles move by changing the barometric pressure setting. See how the accuracy of the altimeter depends on having a correct setting for the barometric pressure. Notice which way the needles move when you set a higher or lower barometric pressure.

8. While you are in the right seat, find the **throttle** and move it forward, saying out loud—to fix it better in your memory—"more power". Bring it back toward you, saying "less power". Feel the shape of the knob or handle, and notice its color (black). Which gauge would show you—if the engine were running—how much you had increased or decreased the power? While you are moving the throttle with one hand, touch that gauge with a finger of the other hand.

9. Does your airplane have a **prop control**? If it does, move it forward as far as it will go, saying "high RPM". Which gauge will show you the RPM? Touch it.

10. Find the **mixture control**. Move it all the way forward, saying "full rich". Notice and feel its odd shape and its color (red).

11. Find the **fuel gauge**. Touch it. It should show empty because it requires electrical power to operate. Ask your pilot friend if you can turn on the master switch and see which gauges come to life.

12. Does this airplane have an **autopilot**? Ask how it can be turned on and how it can be turned off. Find out how it works. Get your pilot friend to demonstrate, the next time you are flying, and learn to use it yourself.

13. You set your altimeter correctly before takeoff at your home airport. While you are flying along at 5,500 feet, you enter an area of low barometric pressure (a "low" on the weather chart), but you don't know about it. What will your true altitude be—5,500 feet? higher? lower?

14. You have been flying along at 10,500 feet, and now the descent to your destination begins. At about 3,000 feet the engine begins to sound rough and peculiar, as though it is about to quit. What is the likely problem, how could it have been prevented, and how can it be solved in an instant?

15. The airplane is flying along at 7,500 feet, trimmed perfectly for level flight. It has been stable like this for the past quarter hour. No one touches anything, yet suddenly a climb begins. Why? What should be done about it?

16. Ask your pilot friend to show you the Owner's Manual for the airplane. See if you can find the answer to the following questions:

    a. What is the fuel capacity and how many hours of normal flight does that represent?

    b. With full fuel, how much weight can the airplane carry, and how many full-sized adults (approximately) would that represent?

    c. What is the highest this airplane can fly?

    d. Using the appropriate performance chart, find out how much runway would be needed to take off from an airport at 3,000 feet elevation on a warm day (temperature 30°C, 86°F), fully loaded, and safely clear a 50-foot building at the end of the runway.

    e. At what speed will this airplane, carrying full flaps, touch down and simultaneously stop flying after the landing approach? Hint: "Stop flying" means the same as "stall".

    f. If the engine is shut down at 8,000 feet and a splendid airport with long and wide runways is 5 NM away, can the airplane be landed there without power? Assume there is no wind.

2

NAVIGATION: WHERE ARE YOU?
WHERE DO YOU GO NEXT?

Navigation is the art of knowing where you are and how to get from there to somewhere else. There are two ways of navigating—by eye and by instruments. Both methods are often used together.

NAVIGATING BY EYE

Navigating by eye is called "pilotage". It is what pilots did before there were radio navigational aids. You use an aeronautical chart—a map—and match up what you see on the chart with what you see from the airplane. The higher you are, the more the terrain looks like the map. The reason is that the higher you are, the farther you can see, and therefore, the more major features are distinguished, while minor features blend into the background.

In Beryl Markham's wonderful book *West With The Night* (North Point Press, San Francisco, 1983), about the early days of aviation in East Africa, she writes as follows in praise of maps: "A map in the hands of a pilot is a testimony of a man's faith in other men; it is a symbol of confidence and trust. It is not like a printed page that bears mere words, ambiguous and artful, and whose most believing reader—even whose author, perhaps—must allow in his mind a recess of doubt . . . A map says to you, 'Read me carefully, follow me closely, doubt me not.' It says 'I am the earth in the palm of your hand. Without me, you are alone and lost.' "

Look at the section of an aeronautical chart in Fig. 2.1 (page 2-7). The scale is 7 NM to the inch, so the 4-inch by 7-inch rectangle depicted here is about 28 NM wide by 49 NM high—approximately

1400 square miles (nautical) of West Virginia countryside. The map tells you a lot if you know how to read it. But first of all, whenever you use an aeronautical chart, turn it to correspond to your direction of flight. In this exercise you will be flying south, so turn Fig. 2.1 upside down. Otherwise, left-right confusion will set in sooner or later. True, it's harder to read words upside down or sideways, but that is easily remedied by twisting the map momentarily as needed. The importance of orienting the map correctly is that it will then match exactly whatever you see from the airplane—ahead, behind, right, and left.

Let's assume you have just taken off from the Morgantown airport (the solid blue circle at top of map, now near the bottom edge as you are holding it). You are heading south. At first, during your climbout, you will see only the airport and its immediate surroundings. Then shortly, looking out through the right window, you will see the city of Morgantown (shown in yellow), with the Monongahela River winding away to the right. Only after you climb a few thousand feet will you get a broad sweeping view of the terrain. Then the pattern of highways, country roads, railroad tracks, streams, and lakes will become apparent; and villages, farms, factories, mines, and power lines will all come into view. By the time you reach 5000 feet, if the visibility is clear, you will be able to see nearly a hundred miles in every direction.

The point of this imaginary exercise is that low down you are too close to see the big pattern, but from high up the view of the earth matches the map, and it is easy to see where you are. The lesson: If a pilot becomes lost, the best thing to do—if no clouds are in the way—is to climb to a higher altitude. An additional reason to climb, which will be discussed later, is that reception of radionavigation signals and radio communications are better the higher you go.

One field, one hill, one mountain, one stream looks pretty much like any other. So don't depend on natural features alone to tell you where you are. If possible, use unmistakable man-made features in conjunction with the natural ones. Look at Morgantown, for example. The built-up part of the city is depicted as a yellow area of a characteristic shape. This shape looks even more distinctive at night, by the pattern of lights, than it does in the daytime. A river (water is always shown in blue) flows through the city and winds away in large loops toward the southwest. The major airport from which we departed is just east of the city. A large highway (two parallel magenta lines, probably a divided interstate freeway) passes close to the eastern edge of the airport. Looking back, behind the left wing of our airplane, we can see that this highway runs from east to west and then turns sharply southwest just before reaching the airport. Finally, a railroad (a black line with cross bars) runs up into the city from the southeast and crosses the freeway just south of the airport.

Such relationships do not change much with time. The pattern of city, airport, highways, river, and railroad defines the location of Morgantown exactly. With a few more features taken into account, we would have a set of landmarks unique for Morgantown—a set that would not be duplicated exactly anywhere in the United States.

I certainly wished for man-made features some years ago on a flight in Northern Arizona. I was heading for a little airstrip at Canyon de Chelly, and a low overcast kept me from climbing high enough to get a broad view of the countryside. I was too low to receive one of the widely separated radionavigation stations in that desert part of the country. So it was a matter of pilotage, pure and simple. The problem was that down there every dirt road, every mesa, every canyon, and every dry wash looks the same. When the calculated time had elapsed, which should have brought me to the right valley, there was a valley all right, but there was no way to tell by the map if it was the right one or not. It was not. My destination never appeared, but it was a man-made structure that straightened me out—a water tower with KEAMS CANYON painted in big letters on its roof. That put me on the map again, and the rest was easy.

You can play an instructive and challenging game while your pilot friend flies. Close your eyes for 10 minutes and let your friend take you in an unknown direction. Then ask for the local chart, and see how long it takes you to identify your position with certainty. "With certainty"—that's the key phrase. If inexperienced pilots get lost, they become rattled. In our example, they might see a river and a town and a railroad and a highway, and decide it must be Morgantown. But closer study would show that the relationships are all wrong. The town of Philippi, for example, south of Morgantown and near the western edge of Fig. 2.1, has all these features nearby, too, but in entirely the wrong places. The airport, shown by a solid magenta circle (magenta means an airport without a control tower), is at the west rather than the east edge of the town. The railroad never enters the town from the east, but only runs north and south between the town and the airport. After a little calm reflection it would be impossible to mistake Philippi for Morgantown.

Now turn Fig. 2.1 right side up. The color coding shows terrain by elevation above sea level in thousand-foot steps—light green to 1000, darker green to 2000, ivory to 3000, light tan to 5000. Not shown on this chart are the higher elevations, such as one would find in the western part of the country—tan to 7000, beige to 9000, rust to 12,000, brown above 12,000. This color scheme lets you picture the terrain in three dimensions on any chart. Here you can see, about 5 NM east of Morgantown, a low ridge running approximately north-south, with three peak elevations marked—3113, 2216, and 2343 feet above sea level. The 3113 elevation is associated with a tower or similar man-made structure (shown as a blue inverted V) that is 517 feet high (the number in parentheses); this could well be a TV transmitter on the mountaintop. The peculiar triangular symbol at the 2343 elevation shows the mountaintop location of a VOR radionavigation transmitter, to be discussed later in the chapter. Closer inspection shows that the ridge is broken into sections by river valleys cutting through—the Cheat River in a valley by itself as it runs down into the Monongahela, and a smaller stream south of the Cheat, cutting through alongside the railroad and a rural highway connecting Masontown (a few miles east of the VOR station) and Morgantown.

Examining the whole map section now, we can see that high terrain lies to the southeast, with peaks as high as 4020 feet at the southern edge. The very large numeral 4 and smaller 5 (in dark blue) near the center of the chart section mean 4 thousand 5 hundred feet above sea level. This is called the **maximum elevation** of the terrain. It is stated conservatively as an altitude that would clear all terrain in the rectangle by at least 100 feet. If you add another 1000 feet, you will have a truly safe altitude. Maximum elevations like this one are shown in the center of every rectangle on every aeronautical chart. In conditions of poor visibility and especially at night, these charted altitudes (plus 1000 feet) make it possible to fly with complete assurance of being safely above any obstruction.

Aeronautical charts are so packed with information that even experienced pilots may not know the meaning of every symbol. Ask your pilot friend to explain the various things on one of the charts they carry in the airplane. Many of the explanations will be found in the special list on the front of every chart. This exercise is a good way to familiarize yourself with all you need to know for navigation by pilotage. With a chart, some experience, and a little patience, you can find your position with certainty at any time.

To actually navigate by pilotage, you need to draw a straight line from where you are (for instance, from your point of departure) to where you're going. Then mark off 10-NM segments with tick marks along this line. For a chart with a scale of 1 inch = 7 NM, as in Fig. 2.1, this would be a tick mark every one and seven sixteenths inches, or very nearly every inch and a half. These tick marks are called "checkpoints" and you will use them to check the progress of your flight. How long should it take you to fly 10 NM? If your aircraft is a typical general aviation model, it might cruise at around 120 k. If there is no wind, you will cover 2 NM per minute, so you'll cross a checkpoint every 5 minutes.

When you draw the course line, it runs in a certain compass direction, which can be determined from the chart. The rectangular grid lines on the chart run true north, south, east, and west, The magnetic compass seeks the north magnetic pole, which is located in the far north of central Canada, so the compass needle points a little east of north in the western states, nearly north in the midwest, and west of north in the eastern states. Your pilot knows how to obtain the magnetic course from the line drawn on the map. One way is to find the true course first, and then add the **magnetic variation** at your location. The simpler way is to use the compass roses on the chart, which surround each VOR station—these are magnetic directions. Ask your pilot friend to show you how it's done.

So the flight begins on a certain heading, which can be monitored on the heading indicator, as you learned in chapter 1. If this heading is flown accurately, and if the heading indicator is set correctly and doesn't precess too much (cf. page 1-25), and if there is no wind, and if your airspeed is exactly 120 k, you will arrive at the first checkpoint precisely 5 minutes after you start out. Note the time you take off, and keep track of the time continuously. Then, at 5 minutes, look around and identify the terrain landmarks. Are you where you ought to be?

Perhaps after a few tick marks have passed, you observe that you're falling steadily behind. Maybe it took 6 minutes instead of 5 to the first checkpoint, a total of 12 minutes to the second, 18 to the third, and so on. Obviously, something is interfering with the 120 k estimate—most likely a headwind. You can compute your actual groundspeed: 30 NM in 18 minutes = 1.67 NM per minute = 100 NM per hour, or 100 k. Such enroute computations are not only fun, they are important for the safety of a flight because if your groundspeed is much slower than you expected, the fuel you are carrying may not suffice to take you all the way to your destination. The sooner a pilot knows about that, the better they can plan alternatives.

A useful number to write down as you start the takeoff roll on every flight is the **fuel-exhausted time**. Let the pilot tell you exactly how many gallons of fuel are aboard, and how many gallons per hour are burned at the airspeed to be flown. Then you can figure how many hours of flight are possible. If you take off at 9:00 a.m. with fuel enough for 4½ hours of flight, you will be on the ground again by 1:30 p.m. at the latest—beyond a doubt, and preferably at an airport. Keeping track of the continuously revised ETA (estimated time of arrival) at your destination and comparing it with the fuel-exhausted time is a useful exercise for the pilot's companion.

It may become evident, after a few checkpoints have gone by, that the airplane is drifting steadily to one side of the course line. Why? There are three possible reasons: The pilot has not been flying an accurate heading by the heading indicator, the heading indicator itself is not correct (probably because of precession), or there is a crosswind. The commonest cause is a crosswind; then a heading correction is needed. Suppose you have drifted to the left, then the wind must be blowing from the right. By changing heading some 5

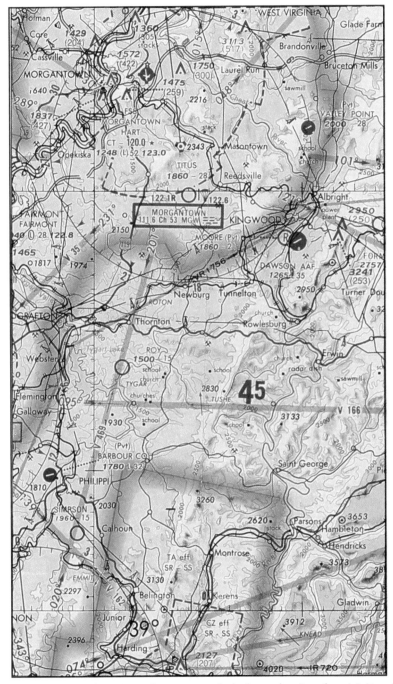

Fig. 2.1. A typical section of an aeronautical chart, showing the vicinity of Morgantown, West Virginia. The scale is 7 NM to the inch, and the figure is a full-size reproduction of the original chart. (Page 2-1)

or 10 degrees to the right (this is called "crabbing" into the wind), the pilot can overcome this drifting tendency. By trial and error and a new correction at each checkpoint, the airplane can be made to track the desired course. Note, however, that every time an airplane crabs, it loses some of its forward groundspeed. It is just as though it flies off course repeatedly into the wind, and then lets the wind bring it back onto course repeatedly—obviously less efficient than flying in a direct straight line.

Now let's go to another part of the country for a sample flight using tick-mark checkpoints. Fig. 2.2 is a chart section that includes Great Falls, Montana. The scale is the same as before. Here the general terrain elevation is above 3000 feet, so the entire chart is a light tan color, but the countryside here is rather flat, so the safe altitude shown by the large blue 4 and smaller 3 is only 4300 feet. Notice that there are two large airports at Great Falls, not to be confused with each other. The one to the east of the city is Malmstrom Air Force Base. The one to the west of the city is the International Airport, used by civilian aircraft. A lot of information about an airport is given right on the chart. Look at as many airports as you can find on the charts, and ask your pilot friend for explanations. At Great Falls, for example, we see that the airport is shown in blue, meaning there is a control tower, a Flight Service Station, or both. The diagram shows three runways at angles to each other. The legend in blue tells us there is both a Flight Service Station (FSS) and a control tower (CT) that operates on radio frequency 118.7. "ATIS" means Automated Terminal Information Service, and it broadcasts on a frequency of 126.6; the use of ATIS is described on page 3-10. The airport elevation is given as 3674 feet above sea level. "L" means there are runway lights. The next number, 105, is the length of the longest runway in hundreds of feet, thus 10,500 feet, or almost two miles. The final number, 122.95, is the "Unicom" frequency; this is described further on page 3-10.

As we did at Morgantown, West Virginia, let's analyze the natural and man-made features that would help a pilot recognize the Great Falls International Airport. Here the Missouri River winds its way in a large loop south of the airport, then turns sharply north to flow through the city. A freeway curves in a horseshoe bend around the eastern edge of the airport, between it and the city.

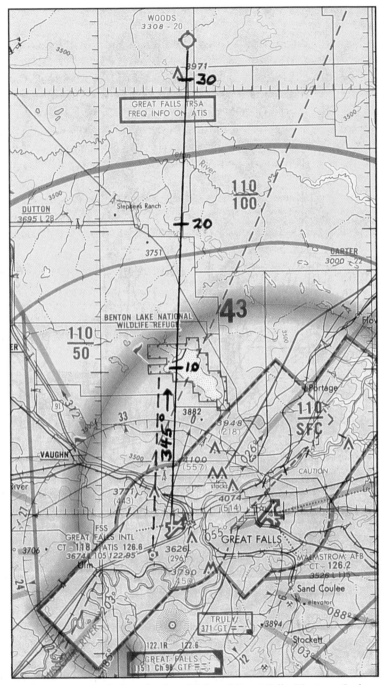

Fig. 2.2. Aeronautical chart for the vicinity of Great Falls, Montana. Scale as in Fig. 2.1. (Page 2-8)

Our flight, for illustrative purposes, will be a short one, from Great Falls to the Woods airport at the very top of the chart. Woods is out in the country. The color is magenta, so there is no control tower. Moreover, the runway is unpaved, as shown by the airport symbol being an open circle rather than a solid one. I drew a straight line between the starting point and the destination, and I marked it off with tickmarks for the 10-mile checkpoints. The total distance is about 33 NM, a 17-minute flight, more or less.

Notice the compass circle around the Great Falls airport. In the center of the circle is a VOR radionavigation facility, the same little triangular symbol that we saw on the Morgantown chart. The numbers around the circle stand for compass directions from the VOR station, beginning with 0 at the arrowhead (meaning magnetic "north"), and going around the whole circle of 360 degrees. As with the heading indicator on an airplane's instrument panel (Fig. 1.21), a zero has been dropped from each number, thus 33 means 330 degrees. I drew a broken line from the VOR station parallel to our proposed course. It cuts the circle at 345 degrees. So 345 degrees is the course we want to fly, and it will be the initial heading after takeoff.

Let's again assume 120 k true airspeed. Then we expect to cross the first 10-NM checkpoint 5 minutes after takeoff. Note the terrain features associated with that checkpoint. It is directly over the western edge of a dry lake bed (which might contain water at certain seasons). Moreover, a large and small lake are just west of the checkpoint. No roads or structures are in the immediate vicinity. If you look down and around and see that you have just crossed a road (roads are magenta), with the lake bed right ahead, you know the airplane is on schedule. Any deviation to either side of the course should be obvious here, and then a small heading correction back toward the course line would be appropriate.

The 20-mile checkpoint (about 10 minutes into the flight) has a very distinctive inverted T pattern of roads close by, and a power line (the thin black line with a tower shown on it near Stephens Ranch) running from northwest to southeast over to the left of the checkpoint. At about 25 NM (12½ minutes) we expect to cross the small Teton River, flowing from left to right with its main

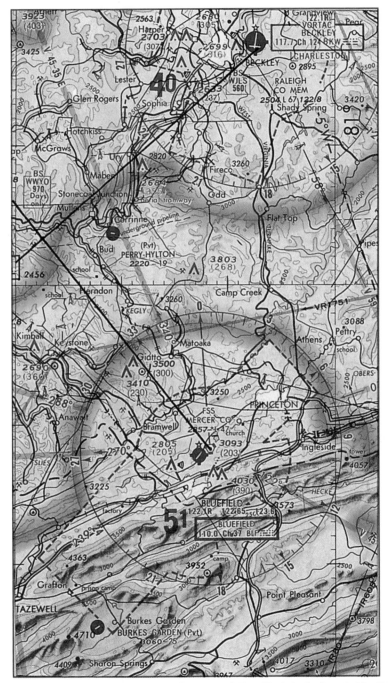

Fig. 2.3. Aeronautical chart for another part of West Virginia. Scale as in Fig. 2.1. (Page 2-20)

tributaries forming a characteristic pattern. Note also that from the 20-mile checkpoint all the way to the destination, a road parallels the course barely half a mile to the right of us.

The 30-mile checkpoint is distinguished by a tower or other structure shown by an inverted V marked with an elevation—3971 feet above sea level. Since the Woods airport, just ahead, is at 3308 feet above sea level (shown by the first number just under the airport name), the obstruction depicted here must be over 600 feet high. Television tower or whatever it is, the structure is marked on the chart for safety—one would want to be well above it, and only descend for landing when it is in full view and can be avoided with certainty.

The best and safest way to find an unfamiliar airport is to remain 2000 or 3000 feet above it and fly right over it. As I explained at the beginning of this chapter, the best view down is from a good way up. It is remarkably easy to miss an airport altogether when scanning ahead from a low altitude. This is especially true when the runway is unpaved, and the worst case of all is a grass strip with green fields all around.

Instructors sometimes feel smug as they see an airport 10 or 15 miles ahead, which is entirely invisible to the student pilot flying the airplane and craning their neck this way and that as they look for it. But it's only a matter of experience in picking out the characteristic clear area that interrupts the normal color and pattern of the fields or wooded areas, Grass or gravel strips really are hard to spot, but an airport with hard-surfaced runways is spread over a lot of real estate. Here's a clue. From a distance, it is much easier if you are looking down the length of a runway than if the runway runs crosswise to your course.

Suppose you have been dozing for a while, and suddenly you wake up and need urgently to know where you are. This might be the case in one of the emergency situations discussed in chapter 5, when you would suddenly have to take over from the pilot. Locating your position will certainly be harder than when you tracked the flight, checkpoint by checkpoint. But just think for a minute! If you know how long you've been flying, and in what direction, you'll

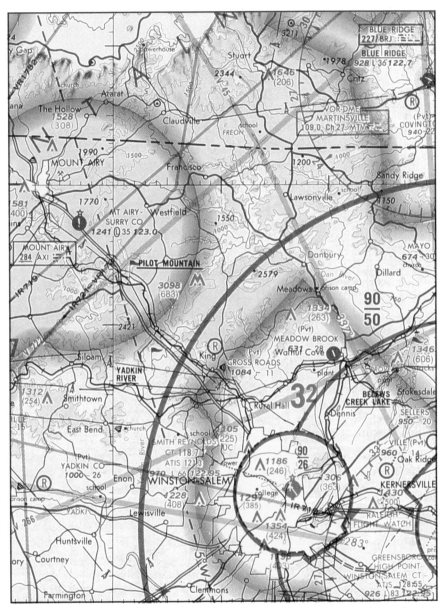

Fig. 2.4. Aeronautical chart for the vicinity of Winston-Salem, North Carolina. Scale as in Fig. 2.1. (Page 2-29, Problems)

have a pretty good idea where you must be. At least you'll know what part of the map to study closely. Here, as with the example I gave about Philippi, West Virginia, you should resist the temptation to jump at a conclusion merely because you see some river or highway or other feature that seems to match something on the chart.

Let's assume you have been flying 13 minutes out of Great Falls toward Woods. You look down and see a nearly-dry lake bed right below. "Aha!", you think, "We must be at the Benton Lake National Wildlife Refuge." But how could that be? Only 10 miles in 13 minutes? Think again. In 13 minutes you probably flew some 26 miles, more or less (or whatever the mileage would be for your airspeed). Your position, therefore, must be somewhere about 26 miles from the Great Falls airport.

It's true that you might have drifted right or left of the course line, so you have to scan across the chart along a 26-mile arc from Great Falls. About 10 miles to the right of the course and about 25 miles out, you will observe that the Teton River forms a kind of lake bed. Notice that at Benton Lake a north-south road runs along the easterly edge of the lake, and the power line crosses this road. Studying the terrain closely at your present position, you see no road and no power line. It looks very much as if the wind has drifted you well to the right of your course, or else an inaccurate heading indicator has taken you there. Anyway, you now have a reasonable idea where you are. Flying northwest from here should bring you to Woods in a few minutes. But don't just fly off in the new direction. Keep track. If you really are at the Teton River, then Woods airport is 2¼ inches, or 16 NM, to the northwest. Make a note of the time, and expect to be over Woods 8 minutes later. If you can't find the Woods airport and its nearby tower after 8 minutes, don't continue. Maintain a safe altitude and circle while you examine the terrain carefully. Then, if you are still lost, turn back south again toward Great Falls. Don't continue blundering on into unknown territory.

The general rule is this: If you have no idea where you are, figure out first how far you might reasonably have come from your starting point in the time that has elapsed, assuming no headwind or tail-wind. Then look around at that distance, sweeping the terrain widely to both sides, until you spot landmarks that match up with the chart. If you can find nothing that looks right, consider that there may have been a headwind or tailwind. A wind of moderate velocity could have slowed you down or speeded you up by 40k (or even more), making your groundspeed 80 k or 160 k instead of 120 k. So extend your scan of the chart accordingly. When you do find what seems to be a good match between the terrain and the chart, look for a more detailed confirmation by checking other prominent natural and man-made features. You can see from this how vitally important it is in every flight to **write down** your time of takeoff. Make that one of your jobs as pilot's companion. This not only helps if you are lost, by telling you exactly how long you have been flying; it also provides the all-important **fuel-exhausted time**, explained above.

NAVIGATING BY INSTRUMENTS

Radionavigation uses the instruments on the airplane to determine geographical position. There are several instruments by which this can be accomplished, and many aircraft have more than one kind aboard. The most important is the VOR system. "VOR" stands for "Very-high-frequency Omni Range". The following explanation will tell you all you need to know about how it works.

Imagine a lighthouse on a small island, with numbers from 0 to 35 painted on the circle of lenses through which the light beam is projected as it rotates. When the beam shines north, it passes through the number 0. When it shines south, it passes through the number 18. As the beam rotates, it passes sequentially through all the numbers (compass directions) until it returns again to north. Now imagine that you are on a ship at sea. When the beam comes around and shines on you, you can look at it through a spy-glass and observe a number. Suppose you see the number 9. You must be somewhere on a line extending out from the lighthouse on a compass bearing of 90 degrees. In other words, you must be due east of the lighthouse.

A VOR station is exactly like the lighthouse. The navigational radio aboard the airplane can "look at" the radio beam as it sweeps past, and thus can tell on what compass bearing from the VOR station the airplane is located. These compass bearings are called "radials" because they radiate out like spokes of a wheel from the VOR station, which is the hub. We are not concerned here with how the electronics work, but only with how to get a practical answer from the system.

What the VOR instrument tells you has nothing to do with which direction your airplane is facing, or in what direction it is flying. The VOR system tells you that you are on a certain radial projecting out from the VOR station. Thus, it gives your direction from the VOR, not how far away you are. That information by itself has some value, and—as we shall see—it can be combined with other information to find the exact location of your aircraft.

To tune in a VOR signal, look first at the chart and find the blue rectangle somewhere near the compass circle that surrounds the VOR station. For example, at the bottom of Fig. 2.2 is the blue rectangle containing the name of the VOR (GREAT FALLS), its frequency (115.1), and its code identifier (GTF) followed by the Morse code equivalent of the identifier (dash dash dot, dash, dot dot dash dot). In order to receive the signal from the GREAT FALLS VOR, the frequency 115.1 must be set on your VOR radio receiver.

The VOR receiver is called a "NAV radio". It is commonly positioned just to the right of the regular communications radio (called "COM radio") on the panel. Wherever it is, you can easily recognize the NAV radio. It may be labelled "NAV", but even if it is not, you can always tell which radio is which by the frequencies that can be selected. NAV frequencies are lower than 118.0, while COM channels are 118.0 or greater.

In Fig. 2.5 the COM radio is at the left, here set to 118.7, the control tower frequency at Great Falls International Airport. The NAV radio, at the right, is set to 115.1, the GREAT FALLS VOR frequency. There is a little switch on the NAV radio with three positions—OFF, VOICE, and IDENT. In order to hear the Morse code identifier, to make sure you are tuned to the right VOR station, you must place this switch in the IDENT position, as in the figure, and the volume control (small knob at lower left, marked "VOL") must be set for an audible signal. In addition, there is sometimes an audio panel—as there is here just above the radios—with a separate switch for directing the output of each radio to the overhead speaker or to headphones. Notice here that the NAV1 switch is in the SPEAKER (up) position so that the VOR identifier can be heard. Different aircraft have different radio arrangements; ask your pilot friend for specific instructions on working the NAV radio.

Fig. 2.6 shows how the VOR indicator works. If you are within range of the VOR, and the NAV frequency is selected on one of the NAV radios, the corresponding indicator comes to life, and the vertical needle swings to one side or another as at the left of the figure. VOR transmissions are like TV transmissions—they travel pretty much by line of sight. This means that at low altitude any hill or mountain between you and the station can block the signal. With a clear line of sight, a typical range would be 50 to 100 NM. All this implies that if the aircraft is low and can't receive any VOR station, the pilot who wants radionavigation guidance should climb, if possible, to a higher altitude.

Fig. 2.5. A COM radio (left), set to communicate with the Great Falls International Airport control tower, and a NAV radio (right), set to receive the GREAT FALLS VOR shown in Fig. 2.2. The large knobs are for selecting frequencies. Volume control knobs are at lower left of each radio. Note audio panel above the radios. See text for further explanation.

By rotating the little knob next to the VOR indicator (sometimes marked "OBS" for "Omni Bearing Selector"), you can make the needle come to the center as at the right of the figure. Note the TO and FROM indications; these are little windows like the one to the right of the needle in the figure. Here the window shows a wedge-shaped pointer directed downward, meaning FROM. As you rotate the knob, you will find that the needle can be centered twice for each complete rotation. At one centering, the TO/FROM indicator will show FROM, at the other centering it will point upward, meaning TO. For the present we want the needle centered in the FROM position, as at right of Fig. 2.6. Then the bearing pointer just above the needle will tell us our compass bearing FROM the VOR (i.e., what radial we are on).

To avoid being misled, it is important to be able to recognize three signs, any one of which warns you that a VOR instrument is not functioning properly.

1. If a red marker (called a "flag") is in view, the system is not working. The flag is located on the face of the instrument, somewhere near the needle. Become

Fig. 2.6.   A VOR indicator. At left, the needle has not yet been centered. At right, the needle was centered by rotating the OBS knob. The TO/FROM wedge marker is showing "FROM" (i.e., pointing down). The bearing indicator at top tells us that this aircraft is on the 045 radial, i.e., directly northeast of the VOR station.

familiar with how it looks by just turning off the NAV radio, or by observing the VOR indicator when the airplane is parked with the switches off.

2. The audible code of dots and dashes is heard if the VOR transmitter is on the air and working properly. If you can't hear the identifier, don't trust the VOR, even if the needle seems to be working! However, make sure first that the volume is turned up fully, and that the correct switch on the audio panel (if there is one) is in the SPEAKER position.

3. A dead VOR instrument behaves dead, the needle doesn't move as you rotate the little knob of the bearing selector. If you do this with a live instrument, the needle will swing from side to side twice during every complete rotation, as described above.

Now let's return to Fig. 2.2. Suppose you are flying somewhere in Montana and you select the GREAT FALLS frequency, 115.1 on your NAV radio. Immediately, the red flag disappears, the needle swings vigorously to one side, and when you turn up the volume and flip the right switches, you hear a Morse code identifier, which should be the same as shown in the box on the chart. The question is: "In what direction are you from the GREAT FALLS VOR?", i.e., on what radial are you located? Rotate the bearing selector knob until FROM is indicated. Some indicator displays have wedges (as in Fig. 2.6), some have windows marked FR and TO, and some display the words TO or FROM with light-emitting diodes.

When the needle centers, verify that FROM is still displayed, as in the down-pointing wedges in Fig. 2.6. If TO is displayed, keep rotating until the needle centers again, with FROM displayed. Now read off from the indicator (or from light-emitting diodes in a direct-reading display) what radial you are on. In some instruments the marker for reading the indicator is at the top, in others it is at the bottom. That is a good example of the crazy lack of standardization we sometimes see in aircraft design. Crazy or not, that's the way it is. So get your pilot friend to show you exactly how to read the particular VOR indicator. When you have the reading, the instrument has told you which rotating beam from the VOR it is receiving.

Let's assume you have centered the needle and confirmed the FROM indication, and you read between "4" and "5" at the top of the dial (as in Fig. 2.6)—i.e., 45 degrees. You must be located somewhere along the 45-degree radial, in other words, on a line that passes from the VOR station through 45 degrees on the compass circle shown in Fig. 2.2. This particular compass circle is labelled only partially, so you have to estimate where 45 degrees would be, half way from 0 (north) to 90 (east). You might be in the vicinity of the Malmstrom Air Force Base, or you might be way off the chart to the northeast. Even though a single VOR only gives information about your direction, knowing what radial you are on can be helpful nonetheless; your search can be confined to a band on each side of a single line on the chart. It may be that—knowing approximately how far from the VOR you are—you can recognize some unmistakable landmark along that line, and thus discover your exact position. However, to locate your exact position by radionavigation alone you need one more piece of information—either the actual distance from the VOR or an intersection of a radial from a different radionavigation facility.

We will consider distance measuring equipment later. Now let's see how to obtain an intersection from another VOR. Look at Fig. 2.3 (page 2-11), a chart of a different part of West Virginia. First, tune in the BECKLEY VOR (117.7) at top of chart, and find (for example) that you are on radial 236; this is drawn onto the chart as a line extending out from the VOR. Now, using the same or another NAV radio, tune in the BLUEFIELD VOR on 110.0, locate what radial you are on, and draw the line. Remember always to verify, after you center the needle, that FROM is displayed. In the example drawn on the chart you are on radial 325 FROM BLUEFIELD. Where that line intersects the other must be your exact position, for it is the only place where you could be on those two radials at the same time.

Is there ever any purpose in centering the needle in the TO position? Yes, indeed. When you do, the bearing pointer will tell you what course will take you directly to the VOR. This course bearing will always be the exact opposite (the reciprocal) of the radial you are on. For example, if you are on the 045 radial (northeast of the VOR), as shown in Fig. 2.6, you would have to follow a course of $045 + 180 = 225$ degrees (i.e., southwest) to reach the VOR.

A simple pencil-and-paper sketch will confirm this if it isn't obvious to you.

In Fig. 2.7 the aircraft is in the same location as in Fig. 2.6, but the VOR indicator has been rotated to center the needle with TO showing. As you can see, the indicated bearing is now 225 degrees, the reciprocal of 045. If you fly that heading, and if there is no wind to blow you off course, you will arrive at the VOR. In the example given in Fig. 2.3, you were on the 325 radial FROM BLUEFIELD. In that case, if you wanted to fly directly TO the BLUEFIELD VOR, you could rotate further, and center the needle again, with the indicator showing TO; the indicated bearing would be 145, i.e., 325 minus 180. Then flying a compass heading of 145 degrees (barring any significant crosswind) would bring you directly to the VOR, right next to the Mercer County airport.

Note that the VOR indicator in Fig. 2.7 has an additional needle—a horizontal one—called a "glide-slope" needle. Its use will be explained in chapter 5. You should ignore it for the present, except to note that there are now two positions for red flags to be displayed—one for each needle. Glide-slope transmitters are only associated with special instrument landing systems (ILS), not with ordinary VOR stations. So a red flag is displayed here as a bright rectangle just at the left end of the horizontal needle, but there is no red flag next to the regular vertical VOR needle. In other words, the glide-slope function is dead but the VOR function is very much alive.

A few words need to be said about a VOR indicator of a different type, called HSI for "horizontal situation indicator" (Fig. 2.8). This

Fig. 2.7.   VOR indicator set for tracking TO a VOR station. The course to the station is 225 degrees. This indicator has a horizontal "glide-slope" needle, which functions only where there is a special "instrument landing system (ILS)".

extremely useful instrument combines the functions of a VOR indicator and a heading indicator. The numbers around the circle are headings, the number under the line at the top is the airplane's heading at the moment—here it is about 070 degrees. Instead of a VOR needle, there is an arrow that rotates as you turn the bearing selector knob. The middle segment of the arrow swings independently to one side or the other; it has the same function as the needle in a standard VOR indicator. Instead of centering a needle, you line up this middle segment with the rest of the arrow, as has been done here. When the arrow is aligned, the little wedge-shaped marker near the center points toward the VOR station, i.e., it points to the compass bearing that would lead you directly to the station. If the arrow is lined up and the marker points away from the arrowhead (as in Fig. 2.8 left), the arrowhead shows the compass bearing FROM the station, i.e., the radial. In this case we are nearly on the 050 radial, our heading is 070, and the VOR station is behind us and off slightly to the right. Most pilots find the HSI more vivid and representational, and a lot easier to use, than an ordinary VOR indicator.

Fig. 2.8. Horizontal Situation Indicator (HSI). At left, the display shows that the aircraft is nearly on the 050 radial FROM the VOR station. Here the wedge-shaped marker points back toward the VOR station. At right, the bearing selector knob has been rotated to center the arrowhead on the reciprocal bearing, 230 degrees TO the station. Now the arrowhead coincides with the wedge marker and both of them point toward the VOR. By combining the HI with the VOR indicator, the HSI presents a vivid and reaslistic picture of the airplane's position.

Just as with the standard VOR indicator display, you can rotate the bearing selector knob here to obtain the reciprocal bearing, i.e., the compass bearing TO the station, in this case 230 degrees, as in Fig. 2.8 on the right. Note that now the marker and the arrowhead point in the same direction, TO the VOR. To fly directly to the VOR, you would turn the airplane to the right, to a heading of 230 degrees. On the completion of the turn, the wedge marker, the arrowhead, and the heading "230" would all be at the top. Then by continuing to fly as the arrow is pointing, just as when flying TO a VOR station with the usual needle display, you would eventually get to the VOR.

For finding your exact position, an easier and more accurate way than by the method of intersections is to have DME (Distance Measuring Equipment) aboard (Fig. 2.9). This device gives a direct readout of the distance in NM from a suitably equipped VOR station. Some VOR stations are not capable of transmitting distance information; those that are have the designation VORTAC, a combination of VOR and the military TACAN, the forerunner of modern DME. Many DME instruments are entirely self-contained, with VORTAC frequency selection and display, as in Fig. 2.9 top, which shows the frequency (here 113.9) as well as the distance (8.9 NM). Others can be connected by a switch to the #1 or #2 NAV, at your option, as in Fig. 2.9 bottom. Note that if radios are stacked one over the other, the one on top is always called #1. The DME in Fig. 2.9 bottom is connected to NAV2, as indicated by the toggle switch at right. The actual ground speed is also displayed—here 71 k—as well as the distance from the station (8.9 NM). As with most radio-navigation equipment, there are many kinds of DME. You should experiment and learn to operate whichever type is installed in your airplane.

Now let's return to Fig. 2.2 and our flight from Great Falls to Woods. We set the bearing selector of a VOR instrument to 345 degrees and tune in the GREAT FALLS VOR (115.1). If we have just taken off from the International Airport, the needle will deviate to the left, because we are to the right of radial 345, just as the course drawn on the chart is parallel to and to the right of the radial. In the present flight, if we decided to fly to Woods along radial 345 (broken line) rather than along the direct course (solid line), we would first take up a heading slightly to the left, say 335 degrees, and wait until the needle centers. That will happen at the moment we reach radial 345, for which the bearing selector is set. Then we would turn back to a heading of 345 degrees and continue to fly outbound on the radial. Then if the needle deviates one way or the other because of wind drift or a heading indicator that has precessed or pilot inattention to heading, a small correction toward the needle will bring us back on course again. Note a simple principle here. The VOR indicator is set to a particular radial. If we are on that radial, the needle will center. If the needle is off to one side, the chosen course is off toward that side. Thus, we speak of "flying toward the needle" to intercept a radial.

Fig. 2.9. Distance Measuring Equipment (DME). Here the frequency of the VOR station being received is 113.9, and the aircraft is 8.9 NM from the station. The upper unit has its own frequency selection, and also has a switch for connecting it to a remote NAV radio ("RMT" is remote position of switch). Somewhere else on the panel there will be a DME switch for choosing whether to connect it to NAV1 or NAV2. This DME is located on the panel directly under the transponder (the unit displaying "1200"), which will be described in chapter 3. The lower unit only operates through NAV1 or NAV2, selected by the toggle switch at the right.

With DME we can watch the miles go by directly. At any moment, by combining DME and VOR information, we can know exactly where we are—on a known radial at a known distance from a VOR. No more wondering if the railroad is on the wrong side of the river! This certainty about where you are, even when you are completely enveloped by clouds and can't see the ground at all, makes real instrument (IFR) flying a precise and safe operation. With proper use of the instruments it is equally impossible to get lost on a visual (VFR) flight, day or night.

We take these marvellous instruments for granted today. It is interesting, while flying along with sure knowledge of one's location, to think back on the early air mail pilots, who had none of these instruments. Their maps were none too good either. Their courses were marked by towers and beacons. If they flew above the clouds, their only means of navigation was by the compass. With no idea what the winds were aloft, it was mostly guesswork, and a wonder that they made it at all. Some of them didn't. If you like the romance of flying in the early times, be sure to read Antoine Saint Exupery's marvellous books *Wind, Sand, and Stars* and *Night Flight.*

Often you don't really care much where you are, you just want to fly directly to a VOR station. This might be the best way to get to an airport in an emergency if (as is frequently the case) there is a VOR on the airport or nearby. Great Falls is a good example (Fig. 2.2); Raleigh County Airport at Beckley, West Virginia is another (Fig. 2.3). Here is the technique for flying directly to a VOR. Find it on the chart, tune it in, make sure no red flag is showing, listen for the identifier (and be certain it's the correct identifier), and double check the frequency against the chart. Center the needle by turning the bearing selector knob, and make sure that TO is displayed. Now the indicated bearing is the course to the VOR, it is the heading you must fly. Turn the airplane gently to that heading, level the wings, and fly. If the needle begins to deviate, make a small heading correction (say 10 degrees) toward the needle. For example, if the needle has moved slightly to the right, turn 10 degrees to the right. This will establish a crab angle into the wind to compensate for wind drift. If, after a while, the needle returns to center and then starts to deviate in the other direction, you probably overcorrected, so make a small heading change (maybe 5 degrees) toward the needle again. Just continue making small corrections as necessary to keep the needle centered, and after each correction, concentrate on keeping the wings level so you will hold a precise heading according to the heading indicator. As you come within a few miles of the VOR station, the needle will become very sensitive, and it will require smaller and more frequent corrections to keep it centered. Eventually, as you come very close to the station, the needle will swing way out, stay there a while, then swing back. When that happens, you are about to cross the station, and when you do, the indicator will flip from TO to FROM.

A final method of radionavigation is by ADF (Automatic Direction Finding). When an ADF radio is tuned to the frequency of a radio transmitter, the arrow of the ADF indicator points directly to the station. Fig. 2.10 shows a typical ADF radio. The band of radiobeacon frequencies is just below the regular broadcast band; the radio in the figure, for example, is tuned to a radiobeacon at 371 kHz. ADF radios can also receive broadcast stations in the usual range 500-1600 kHz. As an aid to pilots, broadcast stations, with their frequencies, are often shown on the aeronautical charts. For example, near the top of Fig. 2.3 is a blue box for station WILS, 560 kHz; and near the left edge of the figure is WWYO, 970 kHz, operating days only. Since broadcast stations are located near cities, and large airports are also found near cities, flying directly to a broadcast station can be a useful way to find an airport.

Fig. 2.10. Typical ADF radio for navigation by radio direction finding. Appropriate frequency (371 kHz) is tuned on the radio, and direction to the station is automatically displayed on the ADF indicator, shown in Fig. 2.11. Note the OFF-ADF-ANT-BFO switch (which must be in the ADF position), the large frequency selector knobs, and the small volume control knob.

An important selector switch on every ADF radio has one position marked "ADF", as shown in Fig. 2.10. Other positions of this switch serve other purposes, but "ADF" **must** be selected to activate the ADF indicator. One type of ADF indicator is shown in Fig. 2.11; another type is much smaller and is built in as an integral part of the ADF radio. The indicator has a little airplane in the center (as on the HI) and an arrow that can point in any direction. Do not be confused by the numbers going around the edge; they have nothing to do with compass bearings or headings, they are simply angles relative to the nose of the airplane. If there is a duplicate set of numbers around the outside, and a knob to rotate them, turn the knob until 0 is straight up, so both sets of numbers coincide, as in the figure. The arrow always points directly to the position of the radiobeacon or broadcast station **relative to the airplane**. If it points straight up (past the little airplane's nose), the station is straight ahead. Straight down past the tail means directly behind. The arrow in Fig. 2.11 is pointing at a radiobeacon located behind and to the left of the airplane.

Since the ADF arrow points at the station, turning the airplane will change it. If you turn the airplane toward the station—a left turn for the example in Fig. 2.11—the arrow, which continues pointing at the station all the while you are turning, will move up toward the nose of the little symbolic airplane. When the real airplane is pointed straight at the station, the arrow will be upright. Note the

Fig. 2.11.   ADF indicator. The arrowhead always points directly to the radiobeacon. Here the station lies along a line extending just behind the left wing. If the airplane turned 120 degrees to the left, the arrow would point straight up, at the nose of the symbolic airplane, and the real airplane could fly directly to the station.

difference between this and the VOR indicator; the VOR needle didn't care which way you faced, it only responded to what radial you were on.

A radiobeacon is shown on the chart as a red stippled circle with a red square nearby giving the name of the beacon, its frequency, and its code identifier. On Fig. 2.2 you can see such a radiobeacon, named TRULY, just south of the Missouri River. Its identifier is GT (dash dash dot, dash). Its frequency is 371, and this is the frequency tuned in on the ADF radio in Fig. 2.10. As pointed out already, the indicator in Fig. 2.11 shows that the station is on a line extending past the left wing tip and slightly to the rear. Notice carefully that with ADF you can't tell where you are except by relating your heading to the ADF needle. Suppose the needle is pointing straight down at the tail of the little airplane, and the heading of the real airplane is due south. Then you must be due south of the radiobeacon, which is directly behind you. On the other hand, with the same ADF indication, if you are heading west, you must be west of the radiobeacon. A little pencil-and-paper sketching of different possibilities should make this perfectly clear.

A story is told, probably apocryphal, about a pilot who started out from an airport in Kansas where there was a radiobeacon on the field. He set his HI 180 degrees wrong, and failed to check it against the magnetic compass. His plan was to fly east to Louisville, Kentucky—about 200 NM. Much to his surprise, he landed at Kansas City, roughly the same distance to the west. He felt perfectly confident in his navigational skills all the way, because he kept the ADF needle pointing straight down during the entire trip! The lesson: Flying to a radiobeacon by keeping the ADF needle straight up is all right, you'll get there. But flying away from a radiobeacon, you depend on the HI to keep you going in the right direction.

Wherever you may be, within reception range, you can tune in a radiobeacon or broadcast station and fly directly to it. This is called "homing". There are more complex ways of getting information about your location from your bearing to a radiobeacon, and you can even find intersections of radiobeacon bearings and VOR radials. In my opinion, however, those methods are not appropriate for the novice. When you are ready, ask your pilot friend to explain those techniques. Homing, on the other hand, is very simple, and it can be a life saver, so it should be learned first.

Suppose you are lost somewhere east of Woods. Tune in 371 on the ADF radio. Listen for the identifier by turning up the ADF volume control and making sure any Phone/Speaker switch marked "ADF" is in the SPEAKER position. Verify that the switch on the ADF radio is in the ADF position. Now turn the airplane very slowly until the ADF arrow points straight up. Then the TRULY radiobeacon will be directly ahead. Note the heading on your HI. This is the heading you must fly. If the arrowhead begins to drift to one side or the other, make small turns toward it, to bring it back upright. Eventually you will cross right over the beacon, and when you do, the ADF arrow will swing from straight up to straight down.

SUMMARY

In this chapter you learned how to navigate by means of the aeronautical chart and by means of the principal radionavigation instruments—the VOR, the DME, and the ADF. There is an old saying among pilots: "I've never been lost, but there have been times when I sure didn't know where I was!" The truth is, you can be confused momentarily about where you are, but by applying what you learned in this chapter, you can always find your position and then—if you wish—fly to the nearest airport. Get your pilot friend to let you practice all these techniques. When you gain some proficiency, you will be able to perform a really useful service from the right seat.

Chapter 2: Problems

(See Appendix for answers.)

1. Fig. 2.4 (page 2-13) shows part of a sectional chart for the vicinity of Winston-Salem, North Carolina. What is the name of the airport at Winston-Salem, what frequency would you tune to hear the current information about conditions there, and then what frequency would you use to communicate with the control tower?

2. For a night flight in this area, what would be an absolutely safe altitude to fly at?

3. What would be the simplest way to fly from Winston-Salem to the Mount Airy-Surry County Airport, about how far is it (same scale as the earlier sectional charts in the chapter), and how could you be sure of finding it?

4. Directly north of Winston-Salem is a high tower, possibly a TV tower. How high is it from base to tip?

5. You are over the town of Mount Airy and you intend to fly directly to the Martinsville VOR. What radial will you be on, and what heading will you fly if there is no significant wind?

6. In the above problem, suppose your VOR receiver were inoperative. Is there any other means of radionavigation that would get you there just as well? Explain how you would proceed.

7. Claudville is a little town northwest of Mount Airy. What landmarks and other aids would tell you for sure that you were over Claudville rather than Danbury, Stuart, Atarat, The Hollow, Francisco, or Westfield? Can you find distinctive and unique characteristics of each of these towns?

8. Under what conditions could you land at the Yadkin County airport?

9. Is there a green and white rotating beacon at the Blue Ridge airport in the northwest corner of the chart? What other airports on the chart have a rotating beacon?

10. This and the subsequent problems are no longer based on Fig. 2.4. Suppose you tune in a VOR, which is shown on your chart as being about 50 NM away. You would like to fly directly to it, because it is along your intended route. However, the red flag on your VOR indicator remains in place, and no signal can be heard. What might the problem be, and what can you do about it?

11. When you rotate the bearing selector, the needle of the VOR indicator centers, the indication is FROM, and the bearing selector shows 45 degrees. What radial are you on? In what general direction must you fly to reach the station? What exact course must you fly?

12. What is the difference between a **course** and a **heading**?

13. You can locate your position by the intersection of two VOR radials or by one VORTAC radial and a DME reading. Which is more accurate?

14. You start flying directly north, and your ADF needle is pointing straight up, as you proceed toward a radiobeacon. There is a very strong wind from the northeast. Of course, you will be slowed down. If you continue to home on the radiobeacon (i.e., keep the needle pointing straight up), what will happen to your heading as shown on the HI or magnetic compass, and from what direction will you be coming when you finally cross the radiobeacon?

3

COMMUNICATION: HOW TO TALK, HOW TO LISTEN

COMMUNICATION TECHNIQUE

Alexander Graham Bell, it is said, had a terrible time getting anyone to talk into his new device. It was novel and strange, and people were afraid of making fools of themselves. Nowadays everyone uses the telephone freely, but radio still intimidates, although people with CB or ham radio experience will have no trouble. I am writing this for the rest of you, to help you get used to communicating by radiotelephone as painlessly as possible.

Is it really any different from talking on the ordinary telephone? Yes, it is, in two respects. First, in most telephone calls you know whom you're talking to, and they know you, so there is a personal interaction that makes for an easy flow of conversation. Radio communication is more like telephoning a business establishment or government office. You want to find out something or tell them something, but the element of personal acquaintanceship is lacking. You simply want to speak to the person in a certain department, who will be able to tell you what you want to know or will be interested in what you have to say. Always remember, it is a PERSON at the other end (except when it's a tape recording!), so an even temper, courtesy, and a friendly tone of voice will get you the best service.

Second, and more important, in telephone calls the conversation flows freely and efficiently because of the ability to interrupt. Most folks don't realize how an ordinary face-to-face or telephone conversation actually goes, and they are amazed to read a word-for-word transcript. Sentences and thoughts are rarely completed, and fragments—a few words here and there—suffice to convey the meaning. If person A begins a sentence, person B soon breaks in to agree or disagree or request clarification before A's sentence is finished. And whoever is the listener at the moment gives continual feedback to the speaker by exclamations—"Oh!", "Sure", "Yeah", "Really?", "Wow!", "How awful!", "I know what you mean", and so on. All this is missing in radiotelephone conversation.

On the radiotelephone you can either speak (transmit) or listen (receive), you can't do both at the same time. The reason is simple. When the radio transmits, the receiver is automatically shut off to protect it; otherwise the high-powered transmitter, using the same antenna, would overload and damage the sensitive receiver. Besides, you would have a dreadful audio feedback squeal as the same sounds came back through the loudspeaker and right into the microphone again.

This "speak-or-listen-but-not-both" format is very distracting until you catch onto it. A certain rhythm is needed. Pick up the mike—a typical one was seen in Fig. 1.23. Press the mike button, say your message, release the button, listen. A common behavior for the novice is to press the button, thus turning on the transmitter, but not be quite sure what to say. The result is that the channel is blocked; no one else can talk or listen. All the while you are pressing the button, your radio is transmitting, whether you say anything or not. It would be like a broadcasting station at a moment of silence—tuning the dial would show you that the particular frequency was "alive" even though nothing was being said. "Unmodulated carrier" is the technical description of this undesirable state of affairs. Sometimes a mike button gets stuck, and then no one else can communicate on that frequency.

The right technique is to think exactly what words you will say, **then** press the button and say them slowly and clearly, then immediately release the button. This takes a little practice but soon it comes naturally. The way you hold the microphone is important, too. It should actually touch your face near your lips so you are talking directly into it, and then you can speak (in fact you must) in an ordinary telephone voice, without shouting. I have noticed that student pilots, up in an airplane and realizing they're talking to someone **way down there**, tend to shout. Don't do that, it will only make your words sound garbled.

On the ordinary telephone you usually identify yourself by name. On the radiotelephone you identify by the "call-sign" of the aircraft, that is, the aircraft identification number. To make it easier for you, this ID is usually displayed right on the instrument panel on the pilot's side; make a point of finding it every time you get into an airplane. In the U.S., all aircraft ID's begin with the letter "N", so we typically omit it from the call-sign, since it wouldn't contribute anything useful. Following the letter "N" is a series of numbers and sometimes letters. Here are some examples of airplanes I have owned at one time or another (always omitting the leading "N"): 56049, 9SE, 719AG, 2038R. Notice, a number comes first, followed by numbers or letters or both.

In radiotelephone communication the reception is often much poorer than we are used to on the regular telephone. Reasons include weather static, low power, excessive distance, obstacles blocking the line-of-sight path of the radio waves, other stations interfering on the same frequency, defective transmitter or receiver, noise in the aircraft cabin, and so on. To try for perfect understanding anyway, we follow certain conventions about speaking. For the number zero, we say "zero" not "oh". And to distinguish between the number "five" and the number "nine" (which can sound alike if reception is poor) we say "five" and "niner", making the latter into a two-syllable word. The names of many letters sound alike when the fidelity of communication is impaired—for example, B, D, P, T, V. To solve this problem we use a standard phonetic alphabet, given in Fig. 3.1. Thus, for 719AG (for example) we would say "seven one niner Alpha Golf". Practice saying the ID of whatever airplane you're in—especially the last three characters—until it comes naturally.

ALPHA	BRAVO	CHARLIE	DELTA	ECHO	FOXTROT	GOLF
HOTEL	INDIA	JULIET	KILO	LIMA	MIKE	NOVEMBER
OCTOBER	PAPA	QUEBEC	ROMEO	SIERRA	TANGO	
UNIFORM	VICTOR	WHISKEY	XRAY	YANKEE	ZULU	

Fig. 3.1   The phonetic alphabet.

Learning the phonetic alphabet is one of the out-of-the-way things that pilots and would-be pilots do. If your name is Robert (or Roberta), spell it out loud: "Romeo October Bravo Echo Romeo Tango (Alpha)". Where do you suppose the word "Roger" came from, meaning "I have received and understood your message"? The answer is an interesting bit of history. During World War II and earlier, an entirely different phonetic alphabet was in use. It went "Able Baker Charlie Dog Easy Fox..." and so on, to "Roger" for R. Now in Morse code, where a great many abbreviations are in use, R always meant "I have received your message." The R is really an abbreviation for "Received". So "Received" became "R", which was "Roger"—and stayed that way, even though it ought to be "Romeo" according to today's phonetic alphabet.

Excuse the excursion into history. The really important point about phonetic alphabets, as about all aspects of communication, is to get information conveyed clearly. Information mostly means **numbers**—distances, airspeeds, fuel-exhausted time, VOR radials, headings to fly, altitudes and altimeter settings, NAV and COM frequencies. So take special pains to make certain that important numbers are understood. If you hear an important number, repeat it back. If you say an important number, get verification that it was heard correctly. By "important" I mean any number that relates in any way to the safety of a flight and a safe arrival at the destination.

To communicate, the first thing to do is to identify and distinguish COM radios and NAV radios. How to do this was explained on page 2-16. Most communications will be entirely on the COM radio, but one useful method employs the NAV radio as a radiotelephone receiver. In that special case you will be transmitting, always on the same frequency (122.1), on the COM radio and listening on a VOR frequency over the NAV radio. Here is how it works. There are Flight Service Stations (FSS) scattered all over the country. Each of these is connected by telephone cables to many VOR stations. The VOR station has a remote transmitter and receiver. Suppose you are flying over a wilderness area but within range of a VOR. If you transmit on 122.1 (COM radio) you will be heard by the nearest FSS through the remote receiver at the VOR site. The FSS reply will be on the particular VOR frequency that is shown on the chart, and you will receive that on your NAV radio. Unfortunately, not all VORs are equipped for this remote communication function, but you can check it out on the chart. Just above the rectangle that tells the VOR name and frequency, some other frequencies may be listed. These are remote communications frequencies for that VOR. "122.1R" indicates that the controlling FSS, wherever it is, will be receiving on 122.1 and transmitting on the VOR frequency. Other frequencies listed in the same place on the chart are for remote communications conducted in the normal way on a single channel, using only the COM radio.

For an example, let's look back at Fig. 2.2, the area around Great Falls, Montana. If you were anywhere within 50 NM or so of Great Falls, you could transmit on 122.1 (COM radio) and receive on 115.1 (NAV radio, Fig. 3.2). You would open the conversation by calling "Great Falls Radio", identifying your aircraft, and saying that you will be listening on the VOR. Always use the name of the VOR in your first call (followed by the word "Radio"), so the FSS will know which of their many remote outlets to use in replying to you. The FSS operator is receiving and transmitting on several different frequencies, so it's useful for them to be reminded that they should transmit on the VOR frequency. Thus, you might say something like this: "Great Falls Radio, this is Cessna seven one niner Alpha Golf, listening on the VOR", or " . . . listening on one one five point one." Alternatively, here, you could talk and listen on 122.6 on your COM radio, since that frequency is also shown above the rectangle on the chart. Even then, it's a good idea to tell the FSS what frequency you're using and your approximate position, thus: "Great Falls Radio, this is Cessna seven one niner Alpha Golf near Great Falls, one two two point six." Let me give you a few tips about the actual handling of the radio equipment. Different aircraft have different makes and models of avionics, so whenever you get into an aircraft, you should familiarize yourself with the radio system. First, identify the COM and NAV radios, as explained already. Play with the knobs to see which ones set the frequencies. Don't be afraid, you can't hurt anything. Sometimes a single knob does all the work, sometimes an outer knob sets the whole numbers and an inner one sets the tenths and hundredths. Some older-style radios only have tenths, they can't be tuned any finer; newer radios tune to hundredths. The very oldest radios just have a mechanical tuning dial, like many inexpensive car and home radios.

Fig. 3.2. Two typical audio panels. In the upper picture, the radios are set for transmission on 122.1 and reception on the Great Falls VOR (115.1). Note that the NAV radio is set for VOICE reception, the NAV1 selector switch on the audio panel is on SPEAKER, the transmit selector switch is on COM1, and the AUTO switch is on SPEAKER. In the lower picture, the NAV radio is in IDENT position to receive the Morse code identifier. Here the audio panel has push buttons instead of toggle switches.

Find the switch that turns the radio on. Often it's the same as the volume control, but not always. Sometimes a single switch turns on COM and NAV, sometimes they're separately controlled. Some aircraft have only one COM and one NAV, some have two of each. Many aircraft have a separate "avionics master switch" somewhere on the panel. Its purpose is to protect the delicate radio circuits against voltage surges when the engine is first started up. If you can't make any radio work, look for the avionics master switch and turn it on.

You should listen through the loudspeaker. All aircraft are equipped for headphones, too. This means that there is some arrangement for switching the radio output to speaker or phones as desired. Find out how that's done in your aircraft. There may be a SPEAKER/PHONE switch somewhere, perhaps on an "audio panel" above the radios. If there's no audio panel, there might be an automatic system that shuts off the speaker when a phone jack is plugged in; in that case, find the phone jack and unplug it.

The audio panel—if there is one—can cause you problems if you don't understand it. Fig. 3.2 shows two kinds. Typically there is a whole set of SPEAKER/PHONE switches, one for each radio or radionavigation receiver. These are toggle switches in the upper unit, pushbutton switches in the lower one. They are labelled COM1, COM2, NAV1, NAV2, ADF, DME, and so on. They allow you to direct the output of any radio to speaker or to earphones. Both audio panels in the figure also have a switch (marked AUTO) for controlling COM output to speaker or phones in conjunction with an automatic transmit selector switch. The AUTO switch and selector are at the left in the upper unit, at the right in the lower one. The transmit selector switch permits easy changing between COM1 and COM2; it determines which of the two radios will be transmitting, and at the same time it activates the corresponding COM receiver. This all sounds more complicated than it really is, and a little experimentation will make it obvious. It's certainly a lot simpler than operating a home VCR! The main point is that in AUTO mode you will normally transmit on whichever COM radio is indicated by the selector switch, and you will automatically receive on that same COM radio.

As an example of the use of the audio panel, let's consider how to set it up for the kind of communication through a VOR that was discussed earlier. Fig. 3.2 top illustrates 122.1 chosen on COM1, and the transmit selector switch at COM1. The AUTO switch is in SPEAKER position, so we will be able to hear any other aircraft that are transmitting on 122.1. NAV1 is set to the Great Falls VOR frequency 115.1, and the VOICE/IDENT switch on the NAV radio is placed in the VOICE position. Finally, the output of NAV1 is directed to the speaker by means of the NAV1 switch on the audio panel. Make sure the volume controls are turned up; otherwise you can call and call—and be heard just fine—but you'll never hear the reply! In Fig. 3.2 bottom, the NAV receiver is tuned to the GREAT FALLS VOR, but the selector switch is in the IDENT position for positive identification of the Morse code signal.

RADIO FREQUENCIES FOR COMMUNICATION

I have explained the use of 122.1 as a universal transmission frequency, with the reply received over the nearest VOR. And I have also explained how you can communicate through a VOR remote receiver/transmitter, using whatever frequencies are listed at the top of the VOR rectangle on the chart. A common FSS frequency is 122.2, and this can be tried even if it is not specifically listed. All the key frequencies mentioned in this chapter are tabulated in Fig. 3.3, which should be a useful reference for you on all flights.

EMERGENCY: 121.5

GROUND CONTROL: 121.6, 121.7, 121.8, 121.9 (and others).

FSS: Flightwatch (weather): 122.0
      Enroute communications: 122.2 or
                            Transmit 122.1/Receive VOR or
                            As indicated on chart.
      Airport advisory: 123.6

UNICOM: 122.7, 122.8, 123.0 (122.95 at tower-controlled airports)

MULTICOM: (Airport advisory if no UNICOM): 122.9

AIR-TO-AIR: (Between aircraft enroute): 122.75

Fig. 3.3. Key radio communication frequencies.

Another important frequency is 122.0, the so-called "Flightwatch". This is a service for giving pilots weather information and for passing along pilot reports. Thus, while enroute, you can find out what conditions are like ahead, as reported by a pilot who was just there. Set 122.0 on your COM radio and listen. If you are flying within radio range of a Flightwatch remote transmitter you can eavesdrop on valuable conversations. Such eavesdropping is a regular part of the system; it is the best way to keep informed and up to date. When you call, give your approximate position, so the Flightwatch operator can know which of many remote transmitters to use. Even if you don't hear anything on the frequency, give them a call, for instance: "Flightwatch, this is Cessna two zero three eight Romeo near Woods, Montana."

A nearly universal FSS frequency is 123.6. Its special purpose is to serve for airport advisory information at the airport where the FSS is located. These FSS locations are indicated on the chart by "FSS", and such an airport is always designated by a blue symbol.

Airport control towers have at least two regular operating frequencies; the one given on the chart is the one to be used for air-to-ground communication, normally within 25 NM or so. You can find this frequency by locating a blue airport on the chart; if there is a control tower, the notation "CT" will be followed by the tower frequency. For example, in Fig. 2.1 you can see that the Morgantown airport has a control tower with frequency 120.0. It also happens to have a FSS on the field, as shown by "FSS" above the airport name. At Great Falls the control tower frequency is 118.7 (Fig. 2.2). At very large airports there may be more than one tower frequency, each serving a different runway.

Another frequency used by a control tower is assigned to the ground controller, who manages the movement of aircraft on the taxiways. Ground control frequencies seem to be something of a secret; they're not found on the charts but only in special publications. Ask your pilot friend about this. The easiest way to find out, of course, is to call on the tower frequency and ask them. Moreover, when you have landed at a tower-controlled airport, you will be told the Ground Control frequency during your landing roll. For example, they might say "Three eight Romeo contact Ground point seven leaving the runway." What does it mean? It means 121.7, that's what. You can see in Fig. 3.3 that several 121 point something frequencies are in use for Ground Control at different airports. If it's a quite different frequency entirely, like 125.0, they'll say it in full.

At airports without a tower or FSS (magenta symbol on the chart) there is often a designated frequency called "Unicom". This is a radio that is operated (when they have time) by some company selling fuel or aircraft services on the field. The existence of Unicom may be indicated by "U" next to the airport name on the chart, but again, the actual frequencies are not readily found. If your pilot friend has the information—which is in a booklet called Airport/Facilities Directory—that's fine; otherwise you can try all the main possibilities (Fig. 3.3) and see if you can get a reply.

Most tower-controlled airports have an "automatic terminal information service" (ATIS) operating continuously on a special frequency. This frequency is shown on the chart along with the airport name. By tuning in the ATIS about 25 NM from a destination airport, you can learn a lot about what's going on there from the regularly updated tape recording—temperature, wind direction and velocity, visibility, which runway is in use, what frequencies to communicate on, and so on.

At extremely busy airports there may be an additional frequuency called "Clearance Delivery" for aircraft to use before they begin to taxi. In that case, departure instructions will be given first on this frequency, then ground control will give the permission to taxi, and finally the permission to take off will be obtained on the tower frequency.

The final universal frequency you should know about is the emergency frequency 121.5. Don't be afraid to use it. "Emergency" doesn't have to mean you're in big trouble—though it's the best frequency to use if you ever are. Details are given in Chapter 5. This frequency is monitored continuously at most aviation facilities, both civilian and military, throughout the country. So it's a good bet for getting attention and a quick reply if anything at all is amiss.

A perfectly legitimate use of 121.5 when there is no dire emergency would be to establish communication if nobody answers on any other frequency. Let me illustrate from a personal experience. I once landed at an airport shown on my chart in magenta, meaning it had no control tower. As soon as I taxied off the runway, however, I saw a structure that was obviously a control tower, and moreover, they were shining a red light beam in my face. The situation called for some explaining, but how to communicate? No tower on my chart, so no tower frequency either! How could I contact them? The answer was 121.5; I made contact immediately, and then asked for their tower frequency. I learned that the tower had only just been opened, and would appear on the next edition of the aeronautical chart. Your pilot friend will probably comment that I should have been familiar with the NOTAMS relevant to my flight; that's absolutely right, and this is a good opportunity for you to learn from your friend what a NOTAM is.

One more communication frequency could be very valuable to you—the so-called "ATC Discrete Sector Frequency". ATC means "Air Traffic Control", the system used chiefly by aircraft on instrument flight plans. ATC has a very wide network of radar sites, so that virtually every nook and cranny of the continental United States is under radar coverage. In addition, ATC remote transmitter/receiver locations pepper the countryside. One radar site with its associated remote radio facilities is called a "discrete sector". The various sectors for a whole region feed into one ATC center. Oakland Center, for example, covers most of northern California and half of Nevada. Chicago Center covers most of Illinois, Iowa, Wisconsin, and Indiana.

Wherever you are flying, if you are high enough, you are probably within range of ATC radar and an ATC sector controller for communications. The problem is: How do you find out the local sector frequency? The answer: If your pilot friend has instrument enroute charts, they will show the sector frequencies. If not, call the nearest FSS (by methods described already), tell them approximately where you are, and ask: "What is the ATC discrete sector frequency here?" There are good reasons to keep in touch with ATC on any flight and to be followed on their radar. This will be discussed fully in Chapter 4.

COMMUNICATIONS FROM TIE-DOWN TO TIE-DOWN

What remains now is to explain the normal and routine purposes of radiotelephone communications at all stages of a flight. When you understand what's being said, and why, you'll be able to assist your pilot friend by taking over most of the communications.

Let's begin at the beginning of a flight. When your airplane is started up and ready to taxi to the runway, if you are at a field with a control tower, the pilot will need to know what runway is in use, the wind direction and velocity, the altimeter setting, and other weather information. If there is an ATIS on the field, you will tune it in first. Since the ATIS is a transcribed broadcast that repeats itself continuously, you can listen to it as many times as you like until you can catch every word and write it all down (if the pilot is patient enough!).

A typical ATIS might sound like this: "Palo Alto Airport information Tango, two two four five Greenwich weather, two thousand overcast, visibility five, temperature six four, dewpoint five one, wind three three zero at one zero, altimeter three zero two four, landing and departing runway three zero, advise on initial contact you have received information Tango." Every time the broadcast is changed, a new letter of the alphabet is assigned to it; this one is "T" (Tango), the previous one was called Sierra, the next one will be Uniform.

The pilot gets a lot of useful information from this. Let's translate. The time of this recording was 22:45 Coordinated Universal Time (formerly called Greenwich Mean Time), which is 5 hours later than Eastern Standard Time, 8 hours later than Pacific Standard Time—thus, 17:45 (5:45 p.m.) EST or 14:45 (2:45 p.m.) PST. Sometimes Coordinated Universal Time is referred to as "Zulu" time—in this case "two two four five Zulu".

There is an overcast sky with the base of the clouds (the "ceiling") 2000 feet above the airport, the visibility is 5 miles, the temperature is 64 degrees Fahrenheit, the dewpoint is 51 degrees. The dewpoint is the temperature at which water vapor will condense, so when the temperature and dewpoint are nearly the same (not the situation here) there is a danger of fog. The wind is 10 miles per hour coming from 330 degrees (i.e., from a bit west of north). The altimeter has to be set to 30.24, the local sea-level barometric pressure (see page 1-32 and Fig. 1.25); perhaps you can set it yourself and have the pilot check it. When the altimeter is set correctly, it should indicate the field elevation quite closely.

The runway in use is 30. Runways are numbered according to the heading you would fly (dropping the final zero, as on the HI) on takeoff or landing, into the prevailing wind. As you can see here, since the wind is blowing from 330 degrees, runway 30 is a logical choice for takeoff and landing. The other end of runway 30 is called runway 12. This confuses novices because it is the very same strip of asphalt or concrete; but the number it is called by tells us in which direction it is being used.

Does the pilot know how to get to runway 30 or do we need directions? If we need help, the ground controller will be glad to oblige, and that's going to be our next contact anyway. The ground controller is the person in the tower who is responsible for the movement of aircraft on the airport. Set your COM radio to the right frequency (inquire on the published tower frequency if necessary), listen to be sure you're not interrupting anyone, press the mike button, and say something like this: "Ground Control, Cessna seven one niner Alpha Golf at Butler Aviation (or wherever you are), taxi with Tango." By telling where you are, you help the controller to spot you from their perch up in the tower so they can direct you and keep you safely out of the way of other aircraft. If you don't know exactly where you are, ask your pilot friend what to say before you push the mike button. By saying the word "taxi" you tell Ground Control what you want to do, and by saying "with Tango" you tell them that you have all the essential information broadcast by the ATIS.

Be prepared for some surprises when you start communicating. Occasionally a controller sounds like a tobacco auctioneer and you can't understand a word. Don't get rattled, just say calmly: "Would you please say again—very very slowly." The standard phrase "Say again" is the way to ask for a repeat.

When you've listened to pilots talking, you may have thought all the special terms like "Say again" and "Roger" and "Affirmative" and "Negative" were just affectations. Not so. For the same purpose that we use the phonetic alphabet—to achieve clarity—we also use standard phrases for the common things we need to say. It makes it easier on both sides of the conversation and it really does make for better mutual understanding. It not only saves time by avoiding a lot of repetition, it also contributes to safety by avoiding misunderstandings.

You should identify your aircraft on every transmission by saying the last three characters in the ID. The controllers talk to dozens of aircraft, so you can't expect them to recognize all the voices. The procedure is to say your type of airplane and the whole ID the first time you talk to someone, and then subsequently to shorten it down to the last three characters—like introducing yourself by your full name but after that using only your nickname.

Well, now you've told Ground Control that you're ready to taxi, so you're expecting them to give you clearance (i.e., permission) to do so. The reply from the ground controller and the subsequent conversation might go like this:

> *Ground Control:* "Cessna seven one niner Alpha Golf taxi runway three zero by taxiway delta."
>
> *You:* "Niner Alpha Golf, could you please give us directions?"
>
> *Ground Control:* "Niner Alpha Golf left turn as you leave your parking spot, right turn at the end of the row... etc."

Don't ever be reluctant to ask for help; we're all paying for the service, whether we use it or not, and besides, most controllers genuinely want to be helpful. Just remember to be courteous and appreciative at all times. Some personal touches do no harm—phrases like "Good morning, Ground Control", or "I know you're busy and I appreciate your help", or just "Have a good day" when you leave one controller for another.

After taxiing, when the pilot has run up the engine and checked the instruments and flight controls, you'll need clearance for takeoff.

Switch to the tower frequency and you'll be in touch with the person who is responsible for the runway and all aircraft in the air within 5 miles and up to 3000 feet above the airport. This controller is standing right next to your friend the ground controller and therefore knows already who you are, and is expecting your call. All you need to say, when the pilot is ready, is "Niner Alpha Golf ready to go". If you have some special request to depart in an unusual manner, this is the time to say so, for example, "Niner Alpha Golf right downwind departure." Ask your pilot friend to explain downwind, crosswind, and overhead-270 departures.

Typical responses to your request for takeoff clearance might be any of the following: "Niner Alpha Golf . . .

. . . is cleared for takeoff."

. . . hold your position!" (or just "Hold!")

. . . taxi up to and hold short of the runway."

. . . number four for departure."

. . . taxi into position and hold."

. . . in sequence."

Only the last two require explanation. To taxi into position and hold means to go onto the runway and be prepared for takeoff clearance, which will follow as soon as the way is clear—e.g., when an aircraft that has just landed gets off the runway or one that has just departed gets far enough away for adequate spacing. The last phrase, " . . . in sequence", means "Be patient, there are several aircraft to land or depart ahead of you, I'll get back to you when your turn comes."

Let's go back to the beginning now and look at the scenario for starting out from a small uncontrolled airport. Then we'll consider how to depart from an extremely busy large airport. At the uncontrolled airport (no control tower, or control tower closed) you're on your own, and you need no permission from anyone; but communications can nevertheless serve a safety purpose by letting pilots alert each other to what they're doing. If there is a FSS on the field, you are supposed to use their airport advisory frequency 123.6. Otherwise use the local Unicom frequency, and if there is no Unicom, use the Multicom frequency 122.9. If there is a tower on the field, but it is closed, use the tower frequency as an advisory frequency. These are uniform prescribed rules, and all pilots know them (or should). You will transmit, but there may be no replies unless another aircraft is around. Your transmission might be: "Hicksville traffic, niner Alpha Golf taxiing to runway three zero," and then "Hicksville traffic, niner Alpha Golf is rolling on runway three zero, left crosswind departure."

At a very busy airport you would first listen to ATIS and then call "clearance delivery" on the appropriate frequency. How will you know the frequency? The ATIS broadcast will tell you; for example, "All aircraft call Clearance Delivery on 125.8 prior to taxi." Here is how your conversation might then proceed:

> *You:* "Clearance Delivery, Cessna seven one niner Alpha Golf with information Tango departing VFR to the east."

> *Clearance Delivery:* "Cessna seven one niner Alpha Golf VFR to the east, depart runway three zero, maintain runway heading to one thousand feet then turn right to heading zero three zero for radar vectors to the airway. Departure control frequency will be 121.3. Squawk five two zero four."

The Clearance Delivery controller wants you to read back what was just told you, to be sure you got it right. This airport has a mix of jumbo jets and little Cessna 152's (and everything in between) arriving and departing all the time; there's no fooling around here, and little tolerance for mistakes. But don't let the busy environment rattle you. Be patient. Ask for slow repeats, and write it all down, inventing a shorthand of your own if necessary. When you ask the controller to speak very slowly, explain that you are a student pilot— you are, in a manner of speaking. And while you're figuring things out, just say "Stand by please, Niner Alpha Golf." "Stand by" is the magic phrase that gets controllers off your back. Especially now, while you're sitting safely in your parking place, there's no rush at all; you have plenty of time to get organized and discuss everything with your friend in the left seat.

Now let's translate the instructions. First, they are confirming your intention to fly in an easterly direction under visual flight rules. You are to take off on runway 30, so your heading as you climb to 1000 feet above the airport will be approximately 300 degrees. Then you are to turn right to 030 degrees, that is a 90-degree turn—60 degrees to 360 (north) and then another 30 degrees. The purpose of all this is to get you moving in the direction of the airway. The radar controller ("Departure Control") will identify you after takeoff and will give you further headings to fly in order to put you on the airway.

The departure control frequency is the frequency you will switch to upon leaving the traffic pattern, instead of remaining in prolonged contact with the tower; this additional controller relieves the tower controller of a lot of work by taking over responsibility for aircraft shortly after they depart the vicinity of the runway.

The word "squawk" refers to a transponder code; ask your pilot friend to explain how a transponder works, and learn how to set codes on it. A typical transponder was shown in Fig. 2.9, "squawk-ing" 1200, the universal VFR code. Unless instructed otherwise, always have the transponder turned on, with this code, and with the altitude-reporting function selected (often marked ALT, as in Fig. 2.9). Your altitude will then be displayed continuously on the controller's radar screen alongside the "blip" of your aircraft. Here you were told to squawk 5204, so set up that code before takeoff. This discrete code will identify your aircraft positively on the screen.

After takeoff, as your pilot sets course and you leave the tower controller's domain, you might wish them a good day, and they'll probably respond with "Niner Alpha Golf have a good flight!". Then, if you have had instructions to contact Departure Control, you'll do that next; set the new frequency and say: "Departure Control, this is Cessna seven one niner Alpha Golf." Finally, when Departure Control has sent you on your way, you'll be on your own at last. Whom to talk to next? If your pilot has filed a VFR flight plan, it has to be opened (activated) after you're under way. For this you should call the nearest FSS. You already know how to do that, either directly on 123.6 or 122.0 or on another frequency listed on the chart, or by the 122.1/VOR system. Listen first, then when the frequency is clear, call them by name—the name of the place where the FSS is located or the name of the VOR through which you are communicating. It might go like this:

> *You:* "Oakland radio, Cessna seven one niner Alpha Golf, over."
>
> *Oakland Radio:* "Cessna seven one niner Alpha Golf, Oakland Radio, go ahead."
>
> *You:* "Niner Alpha Golf, please open our flight plan, San Jose to Reno, we departed at sixteen after the hour."
>
> *Oakland Radio:* "Roger, niner Alpha Golf, your flight plan is opened."

Notice the correct use of "Roger" here, meaning "I have received and understood your message." "Roger" is not a correct way to say "yes". The word for "yes" is "affirmative".

This is a good time to emphasize how important it is to speak **concisely**. The radio frequencies become very cluttered, and there is often someone waiting to say something more important than what you are saying. So work at keeping it short. Whatever tells the story correctly and clearly is sufficient, even it is only a word. If you go over the sample communications in this chapter, you will see how the very minimum of verbiage is used. Of course, this admonition doesn't apply to an emergency, when you should say whatever you need to in whatever manner suits you best.

Incidentally, one of the biggest favors you can ever do for your pilot friend is to remind them to close their flight plan when the trip is over. Ask your friend to tell you what will happen if that's forgotten. It's not a happy story. My favorite and foolproof way of remembering is to move my wristwatch to the opposite wrist whenever I open a flight plan, and to leave it there until the flight plan is closed. You'll be surprised at how often you glance at your empty wrist in the course of any hour. If you can invent a better system, go ahead. Speaking of dire penalties, one of the very worst things a pilot can do is to forget to put the gear down before landing. Your pilot friend should be appreciative if you check the gear-down green light(s) whenever you are on final approach to a runway, and ask politely: "Shouldn't we have a gear-down light here?" if you don't see one. However, first make sure that your aircraft is really a retractable. Anyway, don't be afraid of making a fool of yourself—you just might forestall a disaster some day, because experience has shown there are many circumstances in which pilots have indeed overlooked the gear-down check and landed on the aircraft's belly.

After opening the flight plan, it would be appropriate to switch to 122.0 and listen to the talk on Flightwatch for a while. Then call them; it might go like this (I've put you in a different place and in another airplane):

You: "Flightwatch, Archer five six zero four niner near Colorado Springs."

Flightwatch: "Archer five six zero four niner, Denver Flightwatch, go ahead."

You: "Zero four niner enroute Pueblo to Cheyenne, request weather and pilot reports."

Flightwatch.: "Stand by, zero four niner, we'll see what we have." The specialist on duty is going to consult the latest weather charts, teletype notices, and recent pilot reports. Then: "Zero four niner, Denver Flightwatch."

You: "Go ahead, Zero four niner."

Flightwatch.: "There is a low pressure area on your route north of Denver with ceilings below one thousand feet and visibility less than one mile in rain and snow. The mountains are obscured. A VFR Beech Baron about 30 minutes ahead of you just turned back and landed at Jefferson County."

You: "Thank you, zero four niner."

Flightwatch: "And zero four niner, what are your flight conditions now?" They want a pilot report from you, which they can pass on to other pilots.

You: "Well, we're at eight thousand five hundred feet and we have broken clouds below us, a very smooth ride. It's been like that all the way from Pueblo, zero four niner."

Flightwatch: "Thank you for the pilot report, zero four niner, and have a good flight, Denver Flightwatch out."

Note that Flightwatch only has the job of passing on information; what you do with it is your own business. In this case your pilot friend may decide to stop in the Denver area until the weather improves. Alternatively, if instrument-rated, they may prefer to file an instrument flight plan and then proceed to Cheyenne through the clouds at a safe altitude under the control of an ATC facility.

As you approach your destination, the first step in communications is to tune in ATIS and get the information that the pilot needs in order to plan how the approach and landing will be carried out. About 25 NM away you will call the tower (in some locations Approach Control first, analogous to Departure Control when you started out). After establishing contact, say where you are and what you want:

You: "Zero four niner is about two five miles south, VFR at eight thousand five hundred feet, landing Stapleton with information India."

The reply will instruct you how to fly in order to get into the flow of arriving traffic at the airport. For example:

Approach Control: "Zero four niner, this is Denver Approach, squawk three three one two and ident."

Here you will set your transponder as specified, and then press the "IDENT" button, in order to light up your "blip" on their radar screen.

Approach Control: "Zero four niner, radar contact two zero miles southeast of Stapleton, descend and maintain seven thousand, fly heading two niner zero for radar vectors to the final approach course for runway zero eight right."

This is getting pretty complex! Your pilot friend will be dealing with it, but you can help a lot by writing down the important numbers—altitudes and headings—and keeping the list updated as new directions are given. Your reward will come some day when the pilot turns to you and asks: "Did they clear us to six thousand or was it seven thousand?" and you have the answer, for sure—without a doubt—on your pad of paper. This kind of assistance is really useful to a pilot—even to an old-timer with thousands of hours of flight time—in busy terminal areas, where things can get very very hectic in the left seat. This is one example from among many I shall be discussing in the next chapter—various ways you can bring together what you have learned in the first three chapters and become a helpful pilot's assistant.

SUMMARY

In this chapter you learned about the techniques of effective communication by radiotelephone, and about the various advisors and controllers with whom you may speak in the course of a flight. By applying this knowledge you can improve your understanding of flight procedures, anticipate what is going to be said next, and greatly enhance your usefulness to the pilot.

Chapter 3: Problems

(See Appendix for answers.)

1. Suppose you get into the parked airplane and the pilot starts it up, but then it turns out that both of you have forgotten the Ground Control frequency. What are your best options? (More than one of the following choices may be correct.)

    a. Shut down the engine, get out, and ask someone the Ground Control frequency.

    b. Start taxiing, figuring that they'll see you from the tower and call you.

    c. Call on 121.5 and ask them the Ground Control frequency.

    d. Call on the tower frequency and ask them the Ground Control frequency.

    e. Call on whatever frequency is already set up on the COM radio and see if you get a reply.

    f. Look it up on the Sectional Chart.

    g. Look it up in the Airport/Facility Directory.

2. Suppose your only NAV radio is at the shop being repaired, but the VOR indicator is still in place in the instrument panel. You are flying, in very good weather, about 10 NM from Great Falls, Montana. Is each of the following statements true or false, and why?

    a. You would be flying illegally, because you are not allowed to fly without at least one working NAV radio.

    b. You could tune in the Great Falls VOR at 115.7 on your COM radio, and by flipping the right switch, you could send the signal to the VOR indicator and thus be able to use the Great Falls VOR for radionavigation.

    c. You would be unable to communicate with the FSS at Great Falls, since you could not receive the VOR frequency.

3. You have just reported to a tower controller that you are 15 miles from their airport and that you plan to land there. In reply, they say "Piper seven seven six four Zulu blah-de-blah eight ceiling blah-de-blah downwind." Their whole transmission was reeled off like a shot, and "blah-de-blah" means you just couldn't catch what they said. What should you do?

4. Say the following aircraft ID numbers out loud, using correct phonetic letters and numbers:

    a. 1599B

    b. 806XY

    c. 37L

    d. 6195R

    e. 5QQ

5. Suppose you are flying in a very sparsely settled mountainous part of the country.

    a. If you call on 122.1 and get a reply on the nearest VOR frequency, whom are you talking to, where are they, and how do they know where you are?

b. If you communicate on the discrete sector frequency for the area you are flying in, whom are you talking to, where are they, and how do they know where you are?

c. If you communicate on 122.0, whom are you talking to, where are they, and how do they know where you are?

6. What is the exact meaning of "Taxi into position and hold" and why is this instruction given?

7. Define the following terms and explain what (if anything) you have to do in each case.

a. SQUAWK

b. CLEARED

c. AFFIRMATIVE

d. IDENT

e. RADAR VECTORS

f. UNCONTROLLED FIELD

g. ATIS

h. RUNWAY 9

4

HOW TO HELP YOUR PILOT FRIEND WITHOUT RUINING THE FRIENDSHIP

There is an important Federal Aviation Regulation, 91.3(a), which reads as follows: "The pilot in command of an aircraft is directly responsible for, and is the final authority as to, the operation of that aircraft." What that means is that you should try to give a pilot only what help that pilot desires. If your friend wants to do everything without your help, or even regards your interest with annoyance, you'd better accept that state of affairs, and act the part of the passive passenger. You will be able to follow the flight in a more informed manner, having read chapters 1-3, but you will not be an active participant. Nevertheless, you should study this chapter and the next one to prepare yourself for the remote possibility that your role may change if the pilot should become disabled.

In the usual case, your pilot will be one who welcomes every spark of interest you show, and will eagerly accept any help you offer. Let's talk about how you can best help that kind of pilot. Just remember, you are dealing here with a problem in interpersonal relations. No pilot wants to give up their responsibility and prerogatives as pilot in command—and the regulations wouldn't allow that anyway. Also, no pilot wants to expose their own errors or inadequate knowledge—and we all make errors, and none of us knows everything. Finally, no pilot needs argumentation about how to conduct the flight. Recognize these facts and go about your helping in a considerate way.

As a rule, if you show interest, the pilot's response will be positive. Your best bet is to become involved in the preflight planning, before you ever get into the airplane. Ask the pilot to explain the proposed flight, show you the route on the chart, and let you listen to the weather briefing and the filing of the flight plan. If it is to be a VFR flight, you could mark off the checkpoints along the route, as described in chapter 2, then study them in anticipation of your role as navigator or assistant navigator. Have the pilot explain how (on what frequency) the flight plan will be activated once you are airborne, and write it down. Listen carefully when the pilot asks the FSS on the telephone whether there are any NOTAMS, AIRMETS, or SIGMETS affecting your proposed flight. NOTAM is an acronym for "Notice to Airmen", AIRMET is a meteorological advisory for light aircraft, and SIGMET stands for "Significant Meteorological" advisory. These advisory notices alert you to such things as runway closures, lights out of service, VOR shutdowns, expected moderate or severe turbulence along your route, hazardous thunderstorm or tornado activity, icing conditions in the clouds, and so on.

If it is to be an IFR flight, have the pilot go over the instrument enroute charts with you. Note how the route will go from VOR to VOR (or, with area navigation equipment, from waypoint to waypoint). Learn how the minimum enroute altitudes (MEA) are shown for each route segment on the chart. Have the pilot explain where you will go (to what alternate airport) if you can't land at your destination, or if your fuel is consumed faster than expected because of unpredicted headwinds. On an IFR flight, the pilot will probably double-check you much more frequently and much more fussily than on a VFR flight. Don't be insulted, it's all to the good. Everything in IFR flying is more exacting, so double-checking and triple-checking can only enhance safety.

You probably can't help much in the preflight walk-around, except maybe to untie the ropes. It's basically a one-person operation. But once in the aircraft, going through the checklist is ideally a two-person job. Ask for the checklist, and read it out, item by item. That is the way it's routinely done on air carriers, by vocal challenge and response between the pilot and copilot:

You (reading): Gust lock.

Pilot: Removed.

You: Switches.

Pilot: Off.

You: Gear handle.

Pilot: Down.

You: Master switch.

Pilot (doing it): On.

And so it goes. In reading down the checklist, make sure you keep your place. Go down the list with your index finger, otherwise it is easy to skip an item, and Murphy's Law guarantees that the skipped item will later cause trouble. Beware of interruptions, because that's the commonest cause of skipping an item. Be extra wary of the pilot doing anything out of order; and if the pilot defers anything for later, make a **written** note to yourself, which you can call out at the end of the regular checklist. For this and later purposes, make sure you have a pad of paper and a pen or pencil handy.

Here's an example of how serious trouble can arise. It's a gusty, windy day, so when you read "Gust lock", the pilot replies: "I'll remove it later." But when is "later"? You've passed it on the check list, so there's no sure way to be certain it will ever be checked again. All it takes then is a bit of a rush at takeoff time, and that deferred item could well be forgotten. I chose this for illustration because it just happens to be a potentially lethal mistake. The aircraft would fly off the runway but then its pitch attitude could be uncontrollable, putting it into a very steep climb, followed by a stall and a nose-down crash. That has actually happened, more than once, and although it's hard to believe, it has happened to high-time professional pilots. And there are other potentially lethal omissions, too. It's better to follow the check list exactly.

There are three categories of things to be done on any flight— **aviate, navigate,** and **communicate.** That's why those were the subjects, respectively, of chapters 1, 2, and 3. Most pilots will be enthusiastic about letting you help to navigate, if it is a VFR flight. You will keep busy. There will be a lot of interaction between you and the pilot as discussions arise about where you really are, which way and how hard the wind is blowing, what your groundspeed is, and what you think the estimated time of arrival (ETA) is at the destination or at some intermediate checkpoint. Nothing can go seriously wrong too quickly, and the pilot can always take your job back if you get lost.

During the enroute phase of a VFR flight, the fact is that you can wander back and forth across your planned course line and no harm will be done. But every flight has to end on a particular runway at a particular airport, so at that phase of the flight, there's little leeway for error. I am reminded of the airliner from Japan that landed in San Francisco Bay a few hundred yards short of the runway. Pretty close, you might say, and the error was relatively small, considering how far the aircraft had flown, but it wasn't good enough! So, helping to find the airport is a useful task for the occupant of the right seat. Review chapter 2 on how to spot the right combination of land-marks—both natural and manmade—that will lead you to the airport. You might follow a railroad, river, highway, or power line. You might look for the two peaks that flank the valley in which your destination lies. You might want to overfly the nearby city first, then turn to an easterly heading and continue for 7 NM (3½ minutes at 120 k)—or whatever. The point is to study the chart **in advance** for a variety of things you will want to look for, so that when the pilot arrives in the vicinity of the destination, **you** will be the most knowledgeable person aboard concerning how to go about finding the airport. Be alert to situations where there are two airports close to each other, such as the International Airport and the Malmstrom Air Force Base at Great Falls, Montana (Fig. 2-2); you would not be welcome at the latter!

There is a very special circumstance at the end of every IFR flight when your assistance can add materially to the safety of the flight. An IFR flight ends in an instrument approach by radionavigation. Typically, in such an approach, the aircraft descends to within 200-800 feet of the ground, depending on what type of approach it is. This minimum altitude for descent is specified on every approach chart—the more sophisticated the radionavigation equipment, the lower you can descend while you are still in clouds or fog. At a designated point, if the runway or the approach lights are in sight, the pilot goes ahead and lands; otherwise, full power is added, and the aircraft climbs out again in what is called a "missed approach procedure". IFR approaches are absolutely safe, provided they are carried out precisely according to the approach plate, with no cheating. Tragic last words: "I can almost see it, I'll just go a little lower here".

The ideal arrangement for an IFR approach uses two people. The pilot flies entirely by instruments and never looks out. The person in the right seat peers continuously ahead and to both sides, trying to get a glimpse of the runway when it first becomes visible. The moment the runway or the approach lights are seen with certainty (with certainty!), the person in the right seat says: "Runway in sight." That's the signal for the pilot to look up and prepare for the landing. This is exactly the system used on the airliners, the person in the right seat being the copilot. Always use exactly those words— "Runway in sight"—it's no time for misunderstandings. Accidents have nearly happened because a copilot said "I have it!", meaning he had the runway in sight; but the pilot thought he meant that he was taking control of the airplane (i.e., "I have the aircraft") and let go the yoke.

The way the system works is that if you don't see the runway—and therefore don't say anything—the pilot will simply execute the missed approach procedure routinely, at the appropriate time or place according to the chart. The big advantage of this two-person arrangement is that there is no diversion of the pilot's attention between the instruments and the outside at a critical time in the approach.

How do you actually carry out this important right-seat responsibility? Ask the pilot where you should expect to see the runway— straight ahead or to one side or the other. This will depend on the angle of the approach course to the runway and also on the crab angle (the angle at which the aircraft is turned into the wind) if there is a crosswind. The answer will tell you where to concentrate your attention, but you should extend your scan from side to side anyway. Do **not**, under any circumstances, pass on information about things you see other than the runway, the runway lights, or the bright approach strobes that look like tracer bullets shooting in toward the runway threshold. You may even see the ground perfectly clearly when you look straight down, but that has absolutely nothing to do with making a safe landing, which requires seeing straight ahead. For example, imagine a fog on the ground, which prevents your seeing more than 900 feet ahead—an inadequate forward visibility for a safe IFR landing; but at 800 feet up, you would see the ground quite well. So reports like "I just saw highway 95 pass under us" or "There's the stadium" are positively dangerous because they encourage the pilot to believe that it might be all right to violate the prescribed minimum altitude "by just a little", on the grounds that if you could see highway 95, it must surely be possible to see the airport—if only the aircraft were a little lower. In summary, then, have a clear understanding with the pilot that you will be silent until you see the runway, the runway markings, the runway lights, or the strobe approach lights.

The job most pilots are willing to turn over next, after navigating, is communicating. As you develop presence of mind and an easy way with the microphone, you'll find yourself becoming increasingly useful and increasingly appreciated. Get reports from Flightwatch, and in return give them reports about flight conditions you are experiencing. Get flight following service from ATC routinely on every flight. You can do this by calling on the discrete sector frequency, as described on page 3-11. The ATC controller's first priority is to handle IFR traffic, but if they are not too busy, they will provide you with flight-following VFR traffic advisories. This means you will get timely warnings about other aircraft in your vicinity.

Here is how traffic advisories work. The ATC sector controller has you identified on the radar screen. This identification is provided by the "squawk" you were assigned for your transponder—a discrete code number like 4267. The controller told the computer that 4267 is you—let's say your aircraft ID is 532BC. Then the computer will translate 4267, and display "532BC" next to your radar blip as you move along. Now suppose another blip is moving on the screen in such a way that the two blips are likely to collide. The conroller estimates the angle at which the other aircraft would appear to you, and warns you accordingly, for example, "Traffic at two o'clock, three miles, slow, altitude unknown" or " . . . altitude readout five thousand five hundred not confirmed", or " . . . altitude three thousand."

The position of the traffic is described relative to your course. Imagine a giant clock face on the floor of the airplane. If there were no wind, so that the nose of the airplane would be pointed straight ahead in the direction of flight, 12 o'clock would be right in front of you, 3 o'clock would be straight out along the right wing, 6 o'clock would be behind you, and 9 o'clock would be straight out along the left wing. So you know where to look; 2 o'clock is about 30 degrees ahead of the right wingtip, or about one-third of the way from the wing to the nose. If you are crabbing into a crosswind as you fly along, you will have to remember that the traffic position called out by the controller won't be quite correct for you. For example, if you are crabbed 20 degrees to the left, in order to maintain your course, what the controller calls 12 o'clock will actually be 20 degrees to the right of the nose, or almost at 1 o'clock. Then 2 o'clock will be nearly straight out the right wingtip instead of well in front of it, and so on. The main point is to scan actively in the general vicinity called out by the controller; don't get fixated on one spot.

What was it they said about altitude? If the controller is in touch with the other aircraft and has had confirmation of altitude from its pilot, you will be given the exact altitude. If the aircraft has an operative altitude-reporting transponder but is not in touch with the controller, you will be given the reported altitude with the qualifying phrase "not confirmed". Finally, if the controller has no altitude information at all, you will be told "altitude unknown". Regardless of what you are told, you will be looking vigorously for that aircraft. Usually you will not find it because it is at a different altitude—maybe thousands of feet below or above you. But don't ever relax your search. Some day you'll look at the 2 o'clock position, and there, coming toward you, will be another aircraft; with plenty of warning, you'll point it out to the pilot in ample time for an evasive turn, climb, or descent. One experience like that will firmly establish your value as a pilot's assistant.

When ATC warns you about traffic, your immediate response should be either "Contact, two Bravo Charlie" if you see the other aircraft, or "Looking, two Bravo Charlie" if you don't. Then, if you do catch sight of the traffic, press the mike button and say "Two Bravo Charlie has traffic", to set the controller's mind at ease. Otherwise they will continue to call out warnings as the two blips come closer on their radar screen. At an appropriate time, if you have not spotted the other aircraft, you should say so: "Two Bravo Charlie negative contact the traffic." Again, this is to keep the controller advised so that you, in turn, can get the maximum assistance from them.

What is an "appropriate" time to report not seeing reported traffic? How long should you wait? That depends. If you were told "Traffic twelve o'clock opposite direction three miles fast", you should get into action quickly. A "fast aircraft" could be doing 250 k or faster, so if you are making 180 k, the two of you are closing at a rate of 430 k. That gives you 25 seconds until you meet. Then I would consider an appropriate time to tell the controller "negative contact" would be 5 or 10 seconds, no later. On the other hand, "Traffic four o'clock three miles slow" would have a very different connotation, and you might look around for half a minute or so before reporting your failure to make contact.

Whether or not you are in touch with ATC, the single most important thing you can ever do for your pilot is to maintain a good scan outside, and call out any traffic you see. For this activity it is efficient to use the same clock system. If you see another airplane below you and just to the right of the nose, call out: "Traffic one o'clock low." Traffic at 5, 6, or 7 o'clock is so far to the rear that for most airplanes that would be in your blind spot. Try to make as wide a sweep of the sky as you possibly can while looking for traffic. The only vocabulary you need is "low, high, level" and the clock system. But you have to learn some judgment about calling traffic. At first you'll be inclined to call it out even when it's so far away you can barely see it as a spot on the windscreen, or so far above or below you that you couldn't possibly get near each other. Doing that constantly will only irritate the pilot. You'll have to learn by experience that what matters most is another aircraft seen to be approximately on the horizon (that means it is at your level!) and either growing in size (coming closer) or already too big for comfort. Also important are aircraft below you that are climbing or aircraft above you that are descending. Any airplane you see getting noticeably bigger is potentially a cause for concern.

When the sun is out, an airplane shadow on the ground may be the easiest way to spot one that is in your blind spot—behind or above or directly under or blotted out by your wings. Learn to spot your own shadow as it moves acoss the terrain. Then any other airplane shadow should be called to the pilot's attention immediately, while you initiate a neck-craning search for the intruder. If you are watching your own shadow and it remains alone, you can be reassured that your airplane has no company either.

At night, note how the position lights on the wing tips of another airplane give you information about that airplane's direction of flight. Every airplane has a green light on the right wing, and a red light on the left, following the age-old custom of ships. If you see it in reverse—a green light on your left and a red on your right—that airplane must be coming at you. Tell the pilot at once. If you see only a green, in front of you, you are looking at the other airplane's right side, so it is moving from left to right. If you see only a green off to your left, that aircraft is flying abreast of you in the same general direction. The same logic, in reverse, applies to seeing only a red light. In these cases the important question is whether there is **relative motion**, as described below for daytime collision avoidance. If you see green on the right and red on the left, you are coming up behind the other airplane.

When you do see another airplane, put your finger on it. By that I mean place your fingertip on the window or windscreen right where the target is. Now the question is whether or not it moves away from your finger. If it does not, one of three things must be true (Fig. 4.1): (1) It is flying parallel to you and at the same speed; (2) It is flying away from you at a constant angular bearing; (3) It is coming closer and is on a collision course with you. Obviously, only the third alternative matters. If you don't understand Fig. 4.1, ask your pilot friend to explain why, if another aircraft is coming closer (growing bigger), the absence of **relative** motion means an impending collision unless one of you changes course. Don't confuse relative motion with the motion of the other aircraft against the terrain. Relative motion means relative to **you**, i.e., moving away from your finger. What you need to be worried about is the aircraft that—although it may show motion against the background of the terrain—continues to occupy the same spot in your windscreen or on your window, **all the while it grows larger**.

Don't make the mistake of thinking that the ATC controller will warn you of all conflicting traffic. They will try to, but their first responsibility is elsewhere, to the IFR flights that are under their direct control. So if the controller is very busy (you'll hear it on the radio), you can't count on their undivided attention to **your** radar blip. At times of high work load, the controller may even refuse your request for flight following service, or abruptly discontinue the service. Moreover, there are airplanes out there without radar transponders, and those often can't even be seen on the radar screen. Consequently, you and your pilot friend ought to be doing a good job of looking out for yourselves—controller or no controller—just as fliers used to do years ago, before there was a radar system.

You may be thinking by now that flying is a pretty risky business, with traffic coming at you from all directions. Nothing could be further from the truth. On a typical cross-country flight you are quite likely not to see even a single airplane for hours at a time.

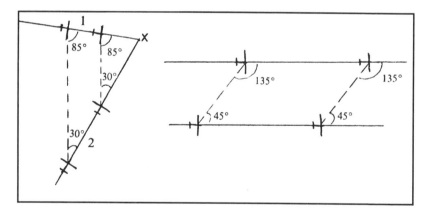

Fig. 4.1. The "fixity of target" principle. At left, two aircraft are on a collision course. Each sees the other at a fixed relative angle, and this remains unchanged. Each sees the other growing larger, until they collide at point X. At right, the relative angles also remain the same, but these aircraft are travelling on parallel courses. The third case is trivial—if the two aircraft at the left were both flying in a direction opposite to that shown, the relative angles would also remain fixed, but they would be getting farther apart (and looking smaller) second by second.

Most of the traffic called out will be at other altitudes. The radar screen covers a lot of territory, so even blips that look to the controller as though they are merging may be hundreds of yards apart in reality. It's a very big sky, and if all the aircraft in the United States were flying at the same time and randomly distributed in the airspace, no pilot would be able to see another aircraft anywhere. The only real problem is in the busy terminal areas and within a couple of thousand feet of the ground near any airport. In those places there will be enough aircraft to make it definitely prudent to keep a watchful eye. Midair collisions are extremely rare, but they have occurred. Take care!

When your pilot friend has gained some confidence in you, you will probably be given an opportunity to aviate, to actually fly the airplane, as described in chapter 1. Unless your friend is a flight instructor, they should not let you land or take off, except possibly to let you participate by holding the yoke very gently while they do the actual work. The way to learn landings (and takeoff, too, if you like) is with a flight instructor who is experienced in "pinch-hitter" instruction.

SUMMARY

In this chapter you learned some ways to apply the techniques of avigating, navigating, and communicating, to be truly helpful to the pilot. You found out some ways to begin practicing these techniques warily, without irritating the pilot. You learned some important specific jobs you can do, such as watching for traffic under VFR conditions, or serving as watchful copilot during IFR approaches. The more you can practice, the more you will gain the pilot's confidence. And the more the pilot learns to depend on you, the more your own self-confidence will grow.

Chapter 4: Problems

(See Appendix for answers.)

1. If there were severe thunderstorm activity forecast for your destination, how would you find out about it?

2. Suppose it is a stifling hot day. You had the door ajar while taxiing to the run-up area, and it remains open during all the pre-takeoff checks. Now your aircraft is cleared for takeoff. How can you help your pilot friend to avoid taking off with the door ajar?

3. My advice about right-seat responsibility during an IFR approach was to remain silent until the runway threshold or runway lights or approach lights are clearly in sight. There is an obvious exception to this rule. What is it?

4. How would you look for traffic at your 6 o'clock position?

5. If you are warned of traffic but don't see it immediately in the direction specified, what should you reply?

6. What is the meaning of a red wing-tip light seen at night off the right side of your airplane and seeming not to move at all?

7. You look down and see the shadows of two aircraft moving across the terrain. What should you do?

5

HOW TO TAKE OVER IN AN EMERGENCY

The purpose of this chapter is to give you confidence that if you had to take over because the pilot was disabled, you could do it. You already know the basics:

1. You know how to keep the airplane flying, wings level, how to make gentle turns, and how to climb and descend.

2. You know how to interpret the most important flight instruments—the altimeter, the airspeed indicator, the attitude indicator, and the heading indicator.

3. You know how to navigate by pilotage and by the VOR system, and you know how to use the ADF to fly directly to a radiobeacon or a broadcasting station.

4. You know how to communicate with control towers and with ATC and flight service stations.

What do I mean by the pilot being disabled? One thinks right away about the worst case, but there are many more likely causes of pilot disability than a fatal heart attack. Here is a simple example. Everything is going along smoothly when the pilot suddenly cries out in pain. It may be nothing more than a bee sting at a sensitive spot— e.g., the eyelid—that could interfere seriously with the pilot's control of the aircraft. If you can take over for just a few minutes, the worst will be over, and the pilot can resume control.

Here is another example: The pilot has a sudden and excruciating abdominal pain and literally doubles over. It could be an attack of food poisoning, or it could be something more serious. Either way, it probably needs medical attention as soon as possible, and until then, the pilot can fly the airplane only with great difficulty. If you can take on most of the work, things will be all right.

Consider a less dramatic but nonetheless serious problem—the pilot suffering from severe fatigue and sleep deprivation, who is also taking sedative medication, and who keeps falling asleep at the yoke. Or the pilot who is frankly under the influence of alcohol or marijuana or some other mind-altering drug. Pilots like that will endanger the lives of everyone aboard unless you insist on taking over.

There is the pilot with blocked Eustachian tubes, who develops excruciating stabbing ear pain during a climb or descent. Air trapped in the nasal sinuses can do the same as it expands or contracts. Unbearable tooth pain can be caused similarly by air trapped in a cavity.

Even in the worst case—a heart attack or stroke—chances are good that quick medical attention can save the pilot's life. Your task, then, is to fly safely to the nearest airport, under instructions from ATC controllers.

WHAT TO DO IMMEDIATELY

Don't panic! What you do in the first minute—even the first 10 seconds—is critically important. Let's suppose you look over to your left and realize, to your horror, that the pilot is slumped over the yoke or slumped back in the seat. If the aircraft is not on autopilot, it will probably already be in an unusual attitude, most likely in a steep bank and diving. The abnormal racing sound of the engine and of the air rushing by the cabin as the airspeed picks up may have been what alerted you to look over at the pilot in the first place.

1. Reach over without delay and pull the pilot away from the yoke.

2. Kill the power by pulling the throttle or power quadrant all the way back in one rapid movement. You don't want engine power making the dive worse!

3. Level the wings immediately, according to what you see outside or on the attitude indicator. If the aircraft is in a steep bank to the left, turn the yoke to the right, and vice versa.

4. As soon as the wings are level or nearly level, pull back gradually on the yoke to bring the little airplane up onto (but not above) the artificial horizon, or to get the nose of the real airplane up toward (but not above) the real horizon, if you can see outside.

5. Start a climb to recover the lost altitude, by pushing the throttle forward to get full power, and do it immediately, as soon as you have returned to level flight. The airplane is probably still trimmed for level flight at a moderate power setting, so full power will automatically start it climbing. It may be very important, because of terrain, that you regain the lost altitude quickly. In that case, turn the trim wheel toward you (or turn the trim crank clockwise) to raise the nose and reduce the airspeed, resulting in a greater rate of climb. However, keep an eye on the airspeed indicator; you don't want the needle anywhere near the bottom of the green arc, where there would be a danger of losing lift altogether.

When an airplane is diving and turning out of control, it can pick up speed very quickly. Unless the procedure described here is carried out right away, the maximum permissible speed (red line on the ASI) could be exceeded. Sooner or later, then, if the airplane doesn't hit the ground first, the speed can pick up enough to tear off the wings. And since an abrupt pull-out from the dive will impose stronger forces than the wings can sustain, pulling back on the yoke to get out of the dive has to be gradual. The best preparation for these critical first few seconds is to practice "recovery from an unusual attitude"—a standard procedure taught to all pilots. This should be part of the regular "pinch-hitter" training (described below), and you should request the instructor to give you this important practice.

Here is a little exercise in reading the instruments. Look at Fig. 5.1 and see how quickly you can grasp the important message the AI and altimeter are telling you. Do it now, before reading further. You should have recognized at once that this aircraft is in a diving turn to the right. You can tell that because the little airplane in the AI is well below the horizon. Moreover, the right wing is down below the horizon, in the dark area, so the airplane is banked to the right at nearly a 30-degree angle. The aircraft is momentarily at 7740 feet above sea level, but it must be losing altitude rapidly, as would be seen clearly if the VSI were shown here. In fact, this is just the situation postulated above, where you must instantly pull back the throttle to kill the power, turn the yoke (here to the left) to level the wings, pull up gradually to stop the dive, then bring in power again to regain the lost altitude.

People often have difficulty remembering which movement of the trim wheel (or trim crank) does what. A good idea is to sit in the airplane on the ground and change the trim, saying "nose up" as you rotate the wheel toward you, "nose down" as you rotate it away. I like to think of looking at the trim wheel from the side, and imagine a model airplane mounted to it, facing forward. Then the rotation

Fig. 5.1. Recognizing an unusual attitude. What's going on here?

that tips up (or down) the nose of the imaginary airplane will do the same for the real one. With a trim crank it's arbitrary—clockwise–nose up, counterclockwise–nose down.

WHAT TO DO NEXT

If you happen to remember what altitude you were at when the trouble began, get back there, then stop the climb by lowering the nose with the trim wheel, reduce the power somewhat, and adjust the trim for level flight at that power setting. If you have no idea what altitude you should be levelling off at, examine the chart and climb 1000 feet above the safe altitude shown there (see page 2-4) for the rectangle you are in. But don't waste time looking for the chart or studying it; just climb, the controller will tell you what to do next.

If the airplane was on autopilot when the trouble began, it is likely that no unusual attitude will have developed, and everything will be easier. The airplane will continue smoothly on its way. You ought to learn the few simple tricks about operating an autopilot. If you usually fly with one, get the pilot to let you practice. In its simplest form, an autopilot is a device that maintains the wings level. To initiate a turn, you have to disconnect it temporarily. Often, however, you can "command" a turn by simply turning a little knob. This gives one a great feeling of power—such a big machine responding to such a tiny manipulation!

More complex autopilots can hold a heading that is set by a little "bug" that you can position on the HI. Then to turn to a new heading, you have only to move the bug. Many autopilots will track a NAV course, i.e., will keep the VOR needle centered, crabbing the airplane as required to stay on the chosen radial. There will also be a selector switch to determine if the autopilot is to work on NAV 1 or NAV 2. Finally, advanced types of autopilot maintain a constant pitch attitude—whatever you set, for level flight, climb, or descent—and also will hold a desired altitude automatically.

If you should be forced suddenly to assume the pilot's job, determine right away if the autopilot is in use, and what it is set to do. You can always shut it down, of course, by flipping a switch or pressing a button, but you're far better off to let it go on working for you until you're on the approach for landing. And don't fight the autopilot. If it is doing what you want, keep your hands off the yoke entirely. If it is not, adjust it to do what you want.

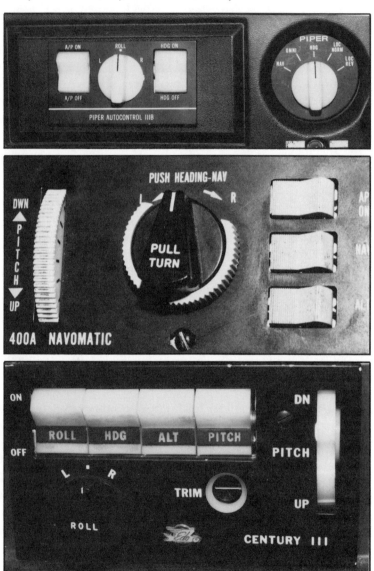

Fig. 5.2. Three makes of autopilot. See text for explanation of the similarities and differences.

Fig. 5.2 shows three popular makes of autopilot. In the one at top you can see an ON/OFF switch, a roll control knob, and a heading ON/OFF switch. The roll control allows you to command a turn to left or right and to roll out level again. The heading switch connects the autopilot to the heading "bug" on the HI. The multiple selector switch lets you track a VOR course. At center is a different model, but the functions are similar. The roll control is the center knob here, which has to be pulled out to disconnect from the heading bug, pushed in again to fly a fixed heading. The autopilot ON/OFF switch is at upper right, a NAV (VOR) tracking switch is at center right, and an altitude hold switch is at bottom right. Pressing the altitude hold switch permits you to hold automatically whatever altitude you are at. To climb or descend, turn that switch off and use the knurled UP/DOWN pitch wheel at left. Finally, the autopilot pictured at bottom has very much the same functions, but the layout is a bit different.

There is only one serious potential problem of which you need to be aware. If an autopilot is in **altitude hold** mode, it will "try" to maintain the same altitude under all conditions. If you reduce the power, it will oppose the natural tendency of the airplane to descend. The only way it can do that is to increase the angle of attack (i.e., raise the nose) and thus reduce the airspeed (see page 1-28). This can get you into serious trouble, because if you keep reducing the power in an attempt to descend, the autopilot will eventually pitch the nose up so steeply that the airplane can not fly at all. So if you do want to descend, be sure to disconnect the altitude-hold function first. And if you ever start a descent by reducing power, and find that the VSI and altimeter don't budge, while the airspeed starts to fall off, suspect an autopilot at once, and deal with it.

With the airplane under control, it's time to get help on the radio. This is very simple, and it will put you in good hands. The two essentials are illustrated in Fig. 5.3.

1. Switch the transponder code from 1200, as it normally is set (see Fig. 2.9), or from whatever code it has been on, to 7700. This simple action rings bells in the ATC center and lights up the blip that is your aircraft on

every controller's radar screen within range. Make sure not only that the transponder is turned on, but that (as in Fig. 5.3) it is in the altitude-reporting (ALT) mode.

2. If you have been in contact with ATC—e.g., if you are on an IFR flight plan or are receiving VFR flight following service—pick up the mike and tell them what has happened.

3. If you have not been in contact with ATC, or you can't get an immediate reply, set your COM radio to 121.5, the universal emergency frequency. If you have two

Fig. 5.3.   Emergency settings of the COM radio and transponder.

COM radios, set them **both** to 121.5. Check the radios carefully, don't get rattled. Make sure they are turned on and that the volume is turned up. Make sure, especially, that the switches marked COM1 and COM2 on the audio panel (if you have one) are both in the SPEAKER position.

4. Press the mike button and say "Mayday, Mayday, Mayday. This is . . ." and give your aircraft ID. "Mayday" is actually the French "M'aidez!", "Help me!"; it is the universal equivalent of SOS in the old Morse code system.

5. Release the mike button and listen, while you pay attention to flying the airplane. Alerted by your transponder code, the ATC controllers were waiting for your call and should answer promptly. From now on, you will receive ample instructions. You will also be asked how much fuel is aboard; if you wrote down a fuel-exhausted time at takeoff (as I recommended), this is what they want to hear now. You can expect to be given headings and altitudes to fly, and you will be brought to a suitable airport. Let me repeat. Once you are in contact with ATC, you can really relax, because they will instruct you, step by step and to the last detail, what you have to do.

6. If you are too low for radar and radio coverage, you may not get a reply. In that case, climb. You can climb safely as high as 12,000 feet (or even 14,000) without oxygen, and then, almost anywhere in the country, you should be within radio range. In some remote mountainous regions of the western states, you may still not be able to communicate effectively. In that case, study the chart and fly to the nearest VOR with voice capability, as described in Chapter 3. Once within range of the VOR, use a frequency shown above the rectangle on the chart. Alternatively, fly toward the nearest airport (preferably one with a control tower or FSS—blue symbol on the chart) and make radio contact with them when you get within range.

LANDING THE AIRPLANE

Now comes the hard part. Your aim will be to fly the airplane onto a runway in such a manner that it touches down in a level or slightly nose-high attitude, not nose-first. You will have been brought to a very large airport (ideally a military one) with an extremely long (a couple of miles) and extremely wide (150-300 feet) runway. An ambulance will be waiting for you, and fire engines will follow you down the runway as you land. All you need to do is set the airplane down somewhere on that vast stretch of asphalt or concrete.

I will describe, step by step, how you should land the airplane, but there is no substitute for a few hours of actual instruction. And a few hours is all it should take to make you competent enough. The basic idea is implicit in the descriptions in Chapter 1 about how to fly the airplane. You will be descending toward the airport at reduced power and an airspeed not too different from what you had been cruising at. This is because you initiated the descent by a simple power reduction without retrimming, and the trim setting determines the airspeed (approximately). But some time before you land, it will be necessary to reduce the airspeed. The slower you can land, the safer, provided the airplane keeps flying until the moment it touches the runway. Here the airspeed indicator is the critical instrument. The aim is to keep the needle well above (say 15 k above) the bottom of the green arc. For most small general aviation aircraft, 70 k would be a good landing speed; heavier and more complex aircraft (including all twins) need more speed, usually landing at 80-100 k.

While descending, you can slow down quite easily by trimming for a more nose-up condition. If you have forgotten how the trim sets the airspeed, refer again to Chapter 1. Suppose you had been cruising at 120 k and now you're descending at 130 k. By trimming more nose-up you can easily reduce the airspeed to 80 k (for example). When you do this, you will find that you are no longer descending so fast, so you will want to make a small power reduction to resume the descent. Your aim is to arrive on final approach to the runway with the airplane trimmed for about the speed at which you expect to touch down. Thus, for small airplanes, 70 k on final approach is about right.

The question whether or not you should turn off the autopilot for your descent and landing is a tricky one. It depends on how familiar you are with the autopilot in your airplane. If you know how to use it, it can be a godsend, because it relieves you of so much work, all the way to touchdown. If you are unfamiliar with it, there could be more problems than it is worth. Here are a few points you should learn about.

The simplest autopilot function is to keep the wings level, and there are some autopilots that do only that, nothing else. Find out how to use your autopilot as a simple wing leveler. Then find out how to make a gentle turn by means of some kind of knob on the autopilot or by temporarily killing the autopilot while you make the turn manually.

Next you need to know how to carry out climbs and descents while the autopilot keeps the wings level. The important thing here is to know how to disconnect any altitude-hold function. As noted already, you wouldn't want to reduce the power while the autopilot held the altitude constant, or the airspeed could fall off to dangerously slow.

If you have practiced these few simple tricks with the autopilot, it can enhance the safety of your descent and landing by ensuring that the wings remain level at all times except when you make small changes of heading. The greatest danger a novice faces in taking over from the pilot is that in the ensuing panic and confusion the airplane will be allowed to enter a steep bank and the dive that is bound to follow. This is what the autopilot protects against. So if you have learned how to use the autopilot effectively, I would recommend keeping it working all the way to your landing. This is contrary to the advice of FAA and the airplane manufacturers, but we are dealing here with a very special situation.

To be able to use the autopilot for landing, you will have to learn one more thing. If your autopilot exerts no control over pitch (nose-up or nose-down attitude), you will simply manage that manually with yoke and pitch trim. However, if the autopilot does control pitch, you have to learn where the little knob or lever is that will raise or lower the nose. Practice that at a safe altitude, because then you will be able to use it at the very last stage of the touchdown. Remember, always, that autopilots are made so that they can be overriden with a little force if necessary. But if it comes to that, you may as well just switch the thing off entirely.

If your airplane has retractable gear, do not forget to lower the gear after these speed reductions have been made. The controller will remind you about the gear, but you may as well do it yourself and take no chances. The gear lever is always clearly marked, but sometimes it can't be moved until some little catch is moved aside, or the lever may have to be pulled out toward you before it can be moved down. Whatever the system is, you will know you have lowered the gear by the green lights that come on—the gear-down lights.

Pilots are taught to use flaps when they land, but I am convinced that a novice like yourself—especially if you have never landed an airplane—should not under any circumstances use flaps for landing. Flaps place the aircraft in a more nose-down attitude for landing. However, when first deployed they may actually make the nose go up, so that the pilot has to trim down to counteract this. The nose-down attitude obtained with full flaps is good for seeing the runway over the nose, but it also makes the actual landing much trickier. Since with flaps the airplane descends more nose-first than without them, an operation called "flaring" has to be carried out at the last minute; flaring means raising the nose slightly just before touchdown. Doing this right is probably the hardest thing in flying, and student pilots go through a lot of bad landings before they learn the feel of it. So, although you will land a little faster without flaps, you will be much closer to the almost-level (slightly nose-up) attitude you need when you touch the runway.

I have spoken glibly about your being on the final approach course to the runway. How will you get there? Don't worry about that, the ATC controller will put you there by a series of easy-to-follow instructions. If possible, they will bring you, by means of radar, to an airport where there are facilities for what is called a PAR, which stands for "Precision Approach Radar". Every military field (and some civilian airports) have this capability—the military call it GCA, "Ground Controlled Approach". When you have established communications with a controller, verify that they do, indeed, plan to bring you to such a facility, and if they have other plans, explain how important a PAR is to you in view of your never having landed an airplane before.

In the PAR/GCA procedure, the radar gives the controller three-dimensional information. It shows not only how well you are lined up on the approach course that leads to the runway, but also whether you are above or below the correct glide path as you descend. There will be a continuous stream of instructions and encouragement, all you have to do is comply. For example:

> "Three eight Romeo start a left turn, now stop the turn . . . that's good . . . now you're a little high . . . that's better, you're coming down to the glide slope now. . . start a right turn, stop the turn, good now, you're doing fine . . . a little low now. . . that's better. . . "

And so on down to the runway.

If the controller says you're a little high, reduce the power very slightly by pulling back the throttle or power quadrant just a few millimeters. You are trimmed for a good airspeed (around 70 k), so don't pull or push the yoke, just use the yoke to start and stop the gentle turns commanded by the controller. If the controller says you're a little low, bring in more power. For obvious reasons, you should worry more about being too low than about being too high.

If you are not being guided by PAR/GCA, you will probably have the help of a VASI, a Visual Approach Slope Indicator. The VASI consists of two pairs of light beams on either side of the runway threshold, so situated that when you are on the glide path, the nearest lights are white and the farthest ones are red. If you get too low, the nearest one will change from white to pink and then to red, so then both pairs of lights will be red. Red means "danger, too low", so add power and get back on the glide path. If you are too high, the farthest lights will change from red to pink to white, so then all the lights will be white. In that case reduce power slightly. The VASI makes it almost as easy as PAR/GCA to come down smoothly to the runway threshold by making very tiny adjustments to the throttle setting.

It is also possible to have radionavigation guidance on your descent toward a runway. If your aircraft is equipped with a horizontal needle (it would be on your #1 NAV VOR indicator, as in Fig. 2.7), you can receive a glide-slope signal. This is part of what is called an ILS (Instrument Landing System). The controller will tell you the frequency and will also talk you down, but the horizontal needle will let you see for yourself if you are too high or too low. You must think of the horizontal needle as your course, in the up-and-down sense, just as the regular VOR needle represents your course in the horizontal sense. In both cases, if the needle drifts, you have to fly toward it to center it again. Here, if you get below the glide path, the needle will drift upward, and you must add more power to get back where you belong. If the needle moves down, you must reduce power because you are too high and you want the airplane to settle down onto the correct glide path again.

Note that the ILS provides the same left-right guidance that one gets from a VOR. With the VOR, one could choose any radial at all and fly to the VOR on it. Here, however, there is only a single course, which leads directly onto the runway. As long as the vertical needle is kept centered, the aircraft will continue inbound to the runway threshold. As long as the glideslope needle is also kept centered, the aircraft will arrive at the threshold at an altitude suitable for landing. Of course, this is easier said than done, and ILS approaches take a long time to learn. But there may be some value in your understanding how they work.

It would be most unusual to have neither PAR/GCA nor VASI nor ILS to guide you, but in that case you will have to manage the descent yourself. Watch the runway threshold. To do this, you may have to raise yourself in your seat to see over the nose. The runway threshold (where the numbers are painted) should remain in the same relative position in your windscreen. If you are too high, the runway threshold will start moving down until it drops out of sight beneath the airplane. If you are too low, the threshold will move steadily up on the windscreen. You must leave the trim alone and not push or pull the yoke. Then you can use more or less power to keep the runway threshold in view in a constant spot as seen through the windscreen. This is really an application of the "fixity of target" principle illustrated in Fig. 4.1, except that here you **want** a collision—so to speak—with the runway threshold. An easy way to remember what to do is to treat the runway threshold exactly like the horizontal glide-slope needle. If it moves down, you are too high, so reduce power. If it moves up, you are too low, so add power.

Throughout your entire landing approach you can concentrate on flying the airplane. You do not have to worry about traffic—the controllers will see to it that there is none. The only tricky part of the landing will be at the very end. Once you are over the runway threshold, with a couple of miles of runway ahead of you, don't be in too much of a hurry to land. Fly along the runway, reducing the power little by little, and bringing the yoke back very very slowly to keep the nose in a slightly high position. The purpose is to fly the airplane onto the runway in as level (or slightly nose-up) an **attitude** as possible. While you were descending, the nose was somewhat lower than the tail, and the nosewheel was lower than the main landing gear. This will not do; landing nosewheel first is a bad idea. So, as you pull the yoke gently toward you, the nose will come up to a level position. Pull just enough on the yoke to keep the nose up; the aim is to keep the nosewheel from striking the ground until the main gear have touched down. If the airplane continues to settle toward the runway, and it seems to be settling too fast, add a little power. If it stops settling altogether, reduce the power slightly. When (at last!) you feel the wheels touch, pull the throttle back briskly to idle.

The instant you fly onto the runway and cut the power completely, you should relax your pull on the yoke—you don't want to take to the air again. Even if you do bounce, just use the yoke to keep the airplane perfectly level or slightly nose-up, and you are bound to settle again. All that remains now, as you roll out on the runway, is to steer by means of the rudder pedals, and to use the toe- or heel-brakes (as the case may be). You should have practiced steering during taxiing at previous flights, so you'll be familiar with how the pedals work—left pedal to go left, right pedal to go right. Stop the airplane as soon as you can, right on the runway, using both brakes together. Pull the mixture control all the way back, to stop the engine. Shut off the magneto switches and the electrical master switch, and open the door. Your job is done. Now it's **their** turn to do what has to be done—to get your pilot to a hospital, to take care of you, and to park the airplane.

Reading realistic descriptions of emergency situations is frightening. I believe, however, that it is a lot better to be frightened needlessly, and never have to use the information, than to put your head in the sand and then—too late—wish you knew what to do, if you unexpectedly need to do something in an emergency. If you believe in being prepared, there is one more practical thing you ought to do. Ask your pilot friend to recommend a flight instructor who will spend a few hours with you in "pinch-hitter" instruction. This means giving you actual hands-on practice handling an airplane, and especially in recovering from unusual attitudes and in landing. Pilot organizations also offer formal pinch-hitter courses that may be useful to you even after studying this book. I hope that you will find pleasant opportunities to use chapters 1-4 in your flying, and that you will enjoy flying more as a result. I hope, too, that you will never have to put this final chapter to use, but that if you ever do, it will prove to be a lifesaver.

SUMMARY

In this chapter you learned how to manage emergencies—especially the emergency in which the pilot becomes incapacitated. By being prepared to get the airplane under control quickly, you can prevent things from getting out of hand. Then, as you have learned, you can get a vast amount of help from the people out there at the other end of the radiotelephone—people who are eager to bring you down safely. You have learned the general methods for landing an airplane. The most important thing to keep in mind is that every year there are several cases in which totally unprepared people in the right seat—who have never flown an airplane and have never read a book like this—have been guided to safe landings. Sometimes their pilots have been able to get medical attention in time, too. Finally, I hope you are stimulated by this book to seek out some actual flight instruction—pinch-hitter instruction if that is all you want, or training for your own license as a private pilot, if you've decided to trade the right seat for the left.

Chapter 5: Problems

(See Appendix for answers.)

1. Your airplane has been on autopilot at 7500 feet above sea level in an area where the charted maximum elevation is 3200 feet. Your heading is 220 degrees. You have not been in radio contract with anyone for the past hour. It is 3 p.m. and the weather is clear. Suddenly your pilot friend slumps over. What should you do and in what order?

2. You are awakened from a nap by a strange rushing sound, and you see at once that the airplane is in a dive. The pilot doesn't respond to your shout. What should you do and in what order?

3. When you are on the landing approach, how can you lower the landing gear, and how can you tell if it is down and locked for the landing?

4. What does it mean if all the VASI lights on the landing approach are red?

5. What does it mean if a glide-slope needle is up toward the top of the indicator rather than horizontal across the middle?

6. How will you know what frequency to set on your #1 NAV in order to receive the glide-slope signal?

7. The airplane is flying about 25 feet above a very long runway and doesn't seem to settle down and land as you would like it to. What should you do?

8. How can you get some actual experience in landing an airplane?

6

APPENDIX

Answers to Chapter 1 Problems

1. a. An airplane without ailerons would be turned with rudder, just as a ship is turned. A ship moves slowly enough, and the water is viscous enough, so that the slow turn works out satisfactorily. If you turn an airplane with rudder alone, it is going so fast and the air is so thin that is skids—i.e., it has so much momentum that it keeps moving in the same direction as before. While it is turning, then, it is also skidding sideways. This is why airplanes are normally turned by banking, but there is no great harm in making a skidding turn. However, it is uncomfortable. At a normal cruising airspeed, ask your pilot friend to show you what a gentle skidding turn is like. You'll feel yourself skidding in your seat toward the outside of the turn.

   b. An airplane without an elevator can easily be made to climb or descend with power changes alone. In fact, this is the exercise you did when you learned how to start a climb by bringing in more power, and how to start a descent by reducing power. If there really were an airplane like this—or if the elevators became stuck somehow—the problem would be how to land it. As you will see in a later chapter, you have to be able to bring the nose up and hold it up as you land, and this can only be done with elevator.

2. Launching gliders by winch is done only where there are updrafts that can carry the glider reliably to a higher altitude. Sometimes there are enough thermals—up-currents of air heated by the sun—to do this. Usually, however, winch launches are done in a valley where some upsweeping air on the upslopes of a mountain range will take over the job, once the glider is airborne.

3. The destination airport is 40 NM to your left. You have flown exactly an hour and have covered 100 NM, and the wind has been blowing you to the right (to the south) all that time. Since the wind is moving at 40 k and you are embedded in the air mass, you have also moved to the south at 40 k.

4. Had you followed the terrain with a sectional chart in hand, you would have noticed very soon that you were drifting to the right, and you would have told the pilot about it. Then the pilot would have crabbed—i.e., turned the airplane a bit into the direction the wind was coming from. In that way, the wind drift could have been compensated for, so that you would arrive directly over the destination airport. Of course, it would take you a little longer than an hour to get there, because you would waste some of your airspeed flying into the wind.

5. There are several reasons you might not be able to go higher than 15,000 feet. **First,** many small aircraft are simply not able to fly that high. The higher you go, the more power it takes just to fly level, because the air becomes thinner and therefore produces less lift. Also, there is less and less oxygen for fuel combustion, so the actual power decreases. Consequently, all aircraft have a "ceiling"—a maximum altitude to which they can fly. **Second,** people need oxygen, too. It would be dangerous for the pilot to fly at that altitude without an oxygen supply, and for the same reason it is actually illegal. The regulations make supplemental oxygen mandatory starting as low as 12,500 feet. **Third,** before climbing up above a cloud layer, it would be very important to know, for sure, that a safe descent would be possible. Therefore, a VFR pilot who does not have an instrument rating that would permit legal flight through clouds, is usually safest not to go on top at all, unless it is certain that the clouds will be broken further along the route.

6. When you bank, some of the lift that was keeping the airplane in level flight is lost. In level flight, all the lift generated by the downwash of air at the wings is directed upward. When the wings are banked, the airplane doesn't "know" it is banked, and the lift is still directed perpendicular to the wings. But this lift is no longer all upward, some of it is sideways. In fact, that's why the airplane is turning—it is being "lifted" toward the center of the turning circle. Since some of the lift has been lost, the result is much the same as if airspeed were reduced (e.g., by a power reduction), and you know already that the nose drops then. It is not something you will be trying, but it is true that when a pilot carries out a steep turn as a maneuver, they have to bring back the yoke—i.e., hold up the

nose—to compensate for the loss of lift. Then as they level the wings again, they have to release the yoke again.

7. a. I assume the AI is tipped over because the gyroscope that runs it is not operating. The gyroscopic instruments work off an engine-driven vacuum pump, which draws air against vanes that spin the gyro. Without seeing your own lifeless AI, I can't say what it is representing, but usually a dead AI shows the little airplane tipped on its side in relation to the artificial horizon. Actually, as you have learned, the airplane is mounted securely to the panel, and the horizon is tipped over. If the little airplane is also against the dark part representing the terrain, it would represent a steep spiral dive—a situation you should never experience unless it was being demonstrated as a maneuver.

b. The markings on the ASI may not go right down to zero, but the ASI at rest will certainly show well below 40 k. For most airplanes this is not an airspeed at which enough lift is generated for flight, so a low reading like that would be seen only in a stall. In aviation a stall has nothing to do with whether the engine is running or not. It means that the wings are not flying, because there is not enough airflow over them. In a stall, the nose drops, just as it does with any loss of lift. Then the airplane picks up airspeed as it dives, and soon it flies again. This sequence of events is due to what is called the "longitudinal stability" of the airplane.

c. The magnetic compass should show the direction correctly or nearly correctly. Every magnetic compass has a little correction card next to it, but the corrections are so small as not to matter. Remember that magnetic north—in most locations—is not the same as true north. If the magnetic compass is telling you something different from what you believe to be true, this may be the problem.

d. The HI is operated gyroscopically, so there is no reason it should tell you the truth when the engine is at rest. Sometimes the HI is operated electrically, but it will not be operative when the master switch is off. So if it happens to agree, it is an accident, probably because it was in agreement when the airplane was last shut down. Find out from your pilot friend which gyro instruments work off an engine-driven vacuum pump and which ones work electrically. In some aircraft, vacuum for the gyros is produced by the airstream rushing through a venturi tube mounted outside.

e. The above discussion applies also to the turn indicator. Of the three gyro instruments—AI, HI, turn indicator—it is customary to have one operate from a different power source than the others. One doesn't want all the eggs in one basket, so to speak. Usually, the AI and HI are engine-driven, while the turn indicator is electrical; but other arrangements may be found.

f. Increasing the barometric pressure setting in the little altimeter window causes the needles to show a higher altitude. Decreasing the setting makes the instrument show a lower altitude. If the altimeter does not show the correct elevation of your airport, the reason is probably that the barometric pressure has changed since the airplane was parked. Suppose the pressure has fallen. That is the equivalent of flying to a higher altitude, where the air pressure is lower, so the altimeter will "think" it is at a higher altitude, and the needles will show an altitude higher than the true airport elevation. Now you find out the new altimeter setting. It will be lower than what is displayed, for the displayed setting is what was correct when the airplane was parked. Now you will enter the new and lower pressure. This will cause the needles to show a lower altitude, namely, the correct elevation of your airport.

8. The power gauge is the RPM indicator in an airplane with fixed-pitch prop, the manifold pressure (MP) indicator in an airplane with controllable prop. Find out which it is in the airplane you fly. Now the RPM gauge will show zero; but the MP gauge, surprisingly, will indicate around 30 inches. With the engine shut down, it is showing atmospheric pressure. Watch it the next time the engine is started. You will see it decrease to a low reading while the engine is idling. Then watch it at the run-up; you'll see it increase as the power is increased. In cruising flight, depending on the airplane, it will probably read in the 20-25 range.

9. Notice the green arc on the RPM gauge. Most airplane instruments have such a green arc to indicate the usual area in which one operates. The RPM gauge will probably show a range from around 2000 to 2600 (20 to 26 as the gauge is marked in hundreds of RPM).

10. This is the control that is pulled back toward you to lean the mixture. When the airplane was parked, that is how the engine was shut

down. Pulling back the mixture control all the way shuts off the fuel flow to the engine, so that only air is drawn into the cylinders, and the engine quits. That is why the mixture control will always be fully pulled back in a parked airplane. This is the control that needs to be pushed in when an airplane descends from a considerable altitude, where a lean mixture was being used because of the reduced oxygen in the thinner air.

11. The fuel gauge is helpful but not necessary. The best pilots don't depend on it, but actually measure (with a dipstick) how much fuel there is at the start of a flight. Then they know exactly how long they can fly before fuel exhaustion, and they plan trips that will never test that limit. If the fuel gauge showed an unusual rate of fuel depletion during a flight, that would be good reason to land as soon as possible; it could be that a fuel leak is draining the tanks. A fuel gauge that suddenly shows empty is probably defective, or the electrical system that operates it has failed.

12. Using the autopilot is an advanced technique that you may enjoy learning. It's fun flying the airplane with a thumb and forefinger when the autopilot is coupled to a "heading bug" on the HI. Sophisticated autopilots not only keep the wings level but also maintain a constant altitude. Whether or not you learn how to use the autopilot, you should know how to turn it on and off. The autopilot is always turned off before landing.

13. On the basis of the earlier explanation about altimeter settings, the answer to this one should be obvious. As you enter the area of low pressure, the altimeter will "think" you are climbing, and it will begin to show a higher altitude. The pilot, however, will maintain an **indicated** altitude of 5,500 feet by descending, without even realizing it. Therefore, in the low-pressure area, the altimeter will continue to indicate 5,500 feet, but the airplane will actually be lower. This is why pilots have the saying "HIGH TO LOW, LOOK OUT BELOW!"

14. The problem is that at altitude the engine was leaned out to compensate for the reduced oxygen. The red mixture control knob was pulled part-way out to reduce the fuel flow to match the reduced oxygen—this is called "leaning the mixture." Now, in the descent, there is insufficient fuel for the increasing amount of oxygen, the mixture is too lean. The solution is simple and instantaneous—push in the mixture control knob. The problem could have been prevented by enriching the mixture gradually throughout the descent.

15. The airplane has entered an updraft. It is still flying level, just as it was before, but the entire air mass is rising and carrying the

airplane with it. Usually nothing needs to be done about it, because after a short time a compensating downdraft will be encountered, which will return you to the original altitude. All the air cannot rise everywhere at once—if it did, it would leave a vacuum down below. So most up and down air currents are fairly local in extent, with updrafts in one place matched by downdrafts somewhere else. However, as wind sweeps over the terrain there are places where you might be flying in a constant updraft or downdraft—for example, updraft on the upslope of a mountain range, downdraft on the downslope. In that case, the airplane does have to be brought back to the desired altitude by appropriate power changes.

16. The answers to these problems are different for every airplane, so I can only offer certain general explanations here. The problems are a good excuse to involve your pilot friend in showing you how the information is to be found and how the calculations are done. The problems are typical of real ones the pilot needs to solve before a safe flight can be undertaken.

   a. There is probably a Table of Specifications on the inside cover of the Owner's Manual. Much of the information is there, such as fuel capacity and fuel consumption in normal cruising flight.

   b. This problem is typical of an important category of problems dealing with Weight and Balance. An airplane is not only limited in the total weight it can carry (fuel plus passengers plus baggage), there is also a limitation on where the weight is carried—on balance—not too far forward nor too far aft. Small aircraft are made for flexibility, so it is not always possible to carry full fuel and a full load of passengers. Sometimes an airplane with four seats can only carry three adults on a long trip with full fuel, whereas four people could enjoy a local sightseeing flight with fuel tanks half empty. If you are sufficiently intrigued, ask your pilot friend to explain weight and balance computations.

   c. Look up "Service Ceiling." That is the highest altitude at which a very low rate of climb is still possible. Practically speaking, in real airplanes (which are never as efficient as when they left the factory), this is the highest possible flight altitude. In truth, you would probably not quite get there in a real aircraft.

   d. There is a chart (probably in the section labeled "Performance"), which lets you look up the answer to this problem, or at least to get a pretty good idea of the answer. The chart itself is interesting; you may be surprised at how much more runway is needed for takeoff at higher elevations and at higher temperatures than at sea level in cool air. Both effects—elevation and temperature—are due thinning of the air. Less air means less oxygen, therefore less engine power. Thinner air means less lift. For both reasons, **performance** suffers. If you have a chance to fly from a high altitude field, you can ask your pilot friend to explain some of the interesting aspects of **density altitude**.

e. Look up the stall speed with full flaps. In a correctly performed landing, the airplane stalls just as it touches down, i.e., at that very moment it reaches the speed at which there is not sufficient airflow over the wings to sustain the airplane in flight. If the speed at touchdown is above the stall, the airplane may well bounce and fly again. On the other hand, if the stall speed is reached before the wheels touch the runway, the airplane will drop suddenly for a hard landing. Getting this just right takes a lot of practice. Perfect landings are something even experienced pilots are proud of—on the rare occasions they happen!

f. The glide performance is usually given in a section of the Owner's Manual called "Emergencies." Typically, it is 10 to 1, meaning that from 6000 feet (almost exactly 1 NM high) a glide of 10 NM is possible, with no wind to consider. There should not be any difficulty making a normal approach and landing without power at the airport 5 NM away.

Answers to Chapter 2 Problems

1. The airport is shown in blue because it has a control tower. Its name is SMITH REYNOLDS, shown in blue just to the left and slightly above the airport. The Automatic Terminal Information Service (ATIS) frequency is 121.3, and the control tower frequency (CT) is 118.7.

2. The just-safe clearance over the highest terrain is shown by the large blue "3" and smaller "2," i.e., 3200 feet above sea level. You can see where the highest point actually is, at Pilot Mountain, where two towers on top of the mountain reach 3098 feet. The "maximum elevation" of 3200 feet allows at least 100 feet over this highest

obstacle. My recommendation was to add 1000 feet, so a truly safe altitude would be 4200 feet.

3. The simplest route is to follow the freeway out of Winston-Salem for about 26 NM. The airport could hardly be missed from the right seat, since it is just to the right of the freeway and a few miles before the sizeable town of Mount Airy. A good idea would be to ask the pilot to fly enough to the left of the freeway so you have a good view of where the airport should be. Sometimes an airport can be in the blind spot under the airplane if you are directly overhead.

4. The top of the tower is at 1334 feet above sea level. The tower is 263 feet high—the number in parentheses.

5. At Mount Airy you will be on the 255° radial. To fly inbound to the VOR you will have to fly the opposite heading, i.e., 255° plus 180° or 255° minus 180°. The latter gives a more sensible answer directly, 075°. However, the other way also works out—435°, and subtracting 360° gives 075°.

6. There is a radiobeacon right in the center of the compass circle surrounding the Martinsville VOR, in other words, located in the same place as the VOR. Its name is BLUE RIDGE, frequency 227, and it is right on the Blue Ridge airport. So you could fly directly to this radiobeacon by just keeping the ADF needle upright.

7. This is a challenging and interesting problem. The distinctive feature about Claudville is its location at a three-way road intersection, where the east-west road crosses a river, so there should be an obvious bridge there. Of the other town listed, only Westfield is at three-way intersection, but it is not in the same relationship to a river. However a river does cross from east to west a mile or two north of Westfield, and this should help to identify it. Danbury is right on a river, as well as being on a highway that runs northwest to southeast. The two towns most difficult to distinguish from each other would probably be Atarat and The Hollow. Both are on the road out of Mount Airy, however, so they could be identified according to which is nearest Mount Airy. For all these towns, radials from the Martinsville VOR would verify the conclusion reached from observation alone.

8. The Yadkin County airport is marked "R," which means "restricted." So the ordinary pilot may not land there except in an emergency or if prior arrangements were made. It would appear, from the presence of a prison camp nearby, that the airport is primarily meant to serve the prison.

9. Yes. A rotating beacon is indicated by a star on top of the airport symbol. Note that the Surry County and Smith Reynolds airports have beacons, but not the private Meadow Brook airport.

10. The usual reason for the situation presented here is so silly you might feel insulted to have it mentioned, yet everyone—even very experienced airline pilots—has done something as silly as this at least once. You thought you set the right frequency, but you got it wrong! Often the problem is a reversal of numbers, like 109.8 instead of 108.9. If the frequency is surely correct and was checked against an up-to-date chart or directory (after all, they do sometimes change a frequency), there is probably obstructing terrain between you and the VOR. Study the chart and decide whether you prefer to continue on course and try again when you get closer, or whether you should climb now to get better reception. Finally, the VOR could be down for servicing or repairs. A call to a flight service station (see Chapter 3) would let you find out if this is the case.

11. You are on the 045° radial, the station is to the southwest, and flying inbound on the 045° radial to the station will require a course of 045° plus 180°, i.e., 225°. You should rotate the bearing selector until TO is displayed and the needle is centered. Then if you keep the needle centered by changing heading slightly as required to compensate for wind drift, you will arrive right over the VOR.

12. A **course** is a track on the chart (i.e., over the ground), designated by its magnetic compass direction. A **heading** is the magnetic direction you must fly in order to stay on a course. When there is no crosswind, they are the same. With a crosswind, the heading will have to be somewhat into the wind in order to prevent drifting off course.

13. The accuracy of an intersection of radials depends on the accuracy of both VOR indicators, each of which typically may be off by a few degrees. This means that locating yourself at an intersection does not really put you at a precise point but rather within a circle or ellipse of uncertainty. Being off as much as a mile would not be at all unusual—how far off depends on how far you are from each VOR, the farther the worse. DME is typically accurate to 1/10 NM, an improvement over the VOR in most situations. However, the uncertainty associated with the one VOR used to determine the radial still remains. On balance, one VOR plus a DME is better than two VOR's, but the best arrangement is to have two VOR's and DME besides.

14. The wind from the northeast, in addition to slowing you down, drifts you steadily toward the west, i.e., to the left. As a result, the needle

will keep moving to the right as it continues pointing at the station. You will be changing heading constantly toward the east in order to keep the needle upright. So your heading, which started out 360° (due north), will be 10°, 20°, and so on. Since the problem did not state how far away the radiobeacon was or how fast you were flying, there is no way to tell how long these heading changes will continue. However, when you are finally flying directly into the wind, there will be no more drift, and you can continue inbound to the radiobeacon on a straight course. As the wind was stated to be from the northeast, 45° will be the most easterly heading you can reach. Then you will be flying directly into the wind, and you will be coming from the southwest as you cross the radiobeacon.

Answers to Chapter 3 Problems

1. We can begin by eliminating a few options. First of all, (a) is foolish, since there are several easy alternatives that don't require shutting down the engine and getting out of the airplane. And (b) is out of the question because it is illegal; at a controlled field you may not taxi without a clearance to do so. You can't do (f) because Sectional Charts do not show Ground Control frequencies.

   The best choice is (g) if there is a copy of the Airport/Facility Directory on board; just look it up under your airport listing, you'll find every local frequency along with a great profusion of other information about your airport.

   The next best choice is (e), it is worth a try anyway. When the airplane was parked, Ground Control was probably the last frequency used before shutdown, so it is probably still set up on one of the COM radios. You can recognize it easily if it is one of the 121 point something possibilities. If calling on that frequency doesn't get a reply, no harm has been done, and you can try something else.

   A straightforward and effective option is (d). After all, you'll be talking to the person in the tower who is standing right next to the ground controller you'd like to talk to. In fact, if things are not very busy, there may well be a single person doing both jobs, and they can give you taxi clearance directly.

   A last resort would be (c). It would be perfectly appropriate if nothing else had worked.

2. a. There is no requirement to have a NAV radio for VFR flight. You do need a COM radio if you fly out of (or into) a tower-controlled airport. Otherwise, you are not required to carry any radio equipment at all.

   b. Wrong. There's no way to receive a VOR frequency on a COM radio—frequencies below 118.0 don't even exist on the COM radio.

   c. Wrong again. Transmitting on 122.1 and receiving on the VOR frequency is only one of many ways to communicate with FSS. You can use any frequency that is shown above the VOR box on the chart—in this case 122.6. Also, near an airport with FSS (as at Great Falls) you can always use 123.6, their universal airport advisory frequency.

3. Say very calmly: "Please say again and speak slowly, six four Zulu." Do this as many times as you need to, without the slightest embarrassment. Their job is to communicate information to you.

4. a. ONE FIVE NINER NINER BRAVO

   b. EIGHT ZERO SIX XRAY YANKEE

   c. THREE SEVEN LIMA

   d. SIX ONE NINER FIVE ROMEO

   e. FIVE QUEBEC QUEBEC (pronounced KAY-BECK)

5. a. You are talking to a FSS through a remote relay at the nearest VOR. The FSS itself could be at an airport anywhere within a range of up to 100 miles or so. They have no idea where you are unless you tell them. However, they sometimes have equipment that allows them to locate you with a direction-finding device; this is called a DF steer, whereby they can give you headings to fly that will bring you to the nearest airport.

   b. You are talking to Air Traffic Control. They are usually a long distance away, in one of the regional ATC centers, which have remote communication relays and remote radar sites. They know where you are both by what you tell them and by identifying you on their radar.

   c. You are talking to Flightwatch. They could be hundreds of miles away. You are communicating through a remote relay. They have no idea where you are except for what you tell them. Their weather and other information for you will be based on where you say you are and where you say you are going.

6. It means you can cross the double yellow line that separates the runway from the taxiway, and you can line up on the runway, ready for takeoff. The instruction is given to save some time. It invariably means that an aircraft has just landed and is on its landing roll before turning off the runway, or that an aircraft has just departed but is not yet far enough away for safe separation from you.

7. a. SQUAWK is the affectionate term for a transponder code, as in "Squawk 3375." Here you would turn the four knobs on the transponder, in turn, to set up 3375 in the transponder window. This gives your radar blip a distinctive unique code, so the controller can follow you along and know who it is.

   b. CLEARED means you have permission to do something for which permission was required. For example, you can be "cleared to taxi" or "cleared to land." You could not be cleared to shut down your engine after you park the airplane. You could not be cleared to do anything at an uncontrolled field.

   c. AFFIRMATIVE means "Yes." It is the only correct way to say it. The reason is supposed to be that "affirmative" and "negative" are easier to distinguish than plain "yes" and "no." I don't believe that, and I'm sure it could be disproved by actual tests, but that's the rule and we follow it. Standardization really does promote safety; the important thing is for **everyone** to behave at all times in precisely the expected manner, following exactly the prescribed rules. No surprises!

   d. IDENT means to push the little button on the transponder that is labeled "IDENT." This changes the pulse pattern of your transponder return in such a way that the blip lights up brightly on the radar screen. Thus, it provides the controller with an additional absolutely positive identification of your aircraft when such identification is needed. It follows that you must never touch the IDENT button unless instructed to do so.

   e. RADAR VECTORS means heading instructions given by a radar controller. In VFR flight such vectors may be obtained by requesting them, if the controller is not too busy with IFR traffic. In IFR flight, radar vectors are mandatory if the controller chooses to use them. Radar vectors can be very helpful in VFR flight in busy airspace because it is a way of being routed around other traffic.

f. UNCONTROLLED FIELD means a field without a control tower, over which ATC exercises no authority. If there happens to be a FSS on the field, they can be a source of useful advisories, and you are supposed to stay in touch with them. But they can not give you clearances, you are on your own. At an uncontrolled field staying clear of other aircraft—taxiing, on the runway, and in the air—is your own pilot's responsibility.

g. ATIS means Automatic Terminal Information Service. It is a continuous broadcast on a special frequency shown on the chart. It is updated periodically and therefore the current one is always identified by a letter—Alpha, Bravo, Charlie, etc. When calling Ground Control for taxi clearance or when calling tower from the air about 15 miles out, you should indicate that you have the ATIS information by giving its letter identifier. Thus, for example, "One Two Three Four Hotel fifteen miles west with November, landing."

h. RUNWAY 9 means that as you took off or landed on this particular runway, your magnetic course would be about 90 degrees, in other words, east. This would be the normal situation if the wind was from the east. This end of the runway would have the number "9" painted on it. The other end would have the number "27."

Answers to Chapter 4 Problems

1. There would be a SIGMET, which your pilot would be told about when the flight plan was filed. The same SIGMET would be read to you if you called Flightwatch on 122.0 enroute.

2. This is an example illustrating why you should make a note to yourself whenever a checklist item is deferred, for whatever reason. Somewhere in the checklist is—or should be—"Doors and Windows." If they are not secured at that time—for good reason in this example because of the heat—just make a note. Then don't let the pilot taxi onto the runway until you have reminded them of any deferred checklist items.

3. If you can see that tree-tops, towers, building, or the like are **dangerously** close, you should certainly warn the pilot, so they can look and see for themselves. It is always remotely possible, though it would be extremely rare, that there is some gross instrument error (or wrongly set instrument), so that the approach is actually unsafe. Of course, you are no expert in judging how close is dangerously close. You'll learn that with enough experience; but meanwhile, better be safe than sorry! If you've given a false alarm, the pilot can look up briefly, be reassured, and return to the instruments again. But if there really is a problem, a pull-up (missed approach) can be executed immediately, and you will have saved the day.

4. With difficulty. This traffic is in the blind spot directly behind you. However, if the pilot will momentarily turn to the right, it should bring the traffic into your field of view to the rear.

5. "(Your ID) looking."

6. It could be another aircraft flying parallel to you and headed in the same direction. It could also be an aircraft on a collision course with you, which is angling toward you and getting closer all the time. This requires warning the pilot and then keeping a very careful watch.

7. Warn the pilot at once, and immediately begin searching in all directions to find the other aircraft. If you don't know which shadow is yours, you can't tell which side to look on, so you need a sweeping search—right, left, front, rear, high, low—until the other aircraft is found.

Answers to Chapter 5 Problems

1. a. Make sure the pilot is not leaning on the yoke.

   b. Check that the airplane is continuing in stable level flight on the autopilot.

   c. Turn the transponder to 7700.

   d. Turn both COM radios to 121.5 and make sure the outputs are directed to the speaker.

   e. Call "Mayday" and give your aircraft ID.

f. When someone replies, explain what has happened and follow instructions.

2. a. Reach for the throttle and bring it back all the way to idle immediately.

d. Level the wings by turning the yoke as necessary, and pull the pilot out of the way if there is interference with yoke movement.

c. When the wings are level, pull gently back on the yoke to bring the airplane out of the dive.

d. When the airplane is level or nearly level, push the throttle all the way in (full power) and climb to a safe altitude.

e. Proceed as in Problem #1.

3. If the aircraft has a retractable gear, there will be a handle marked appropriately. When the handle is operated, and the gear comes down and locks, one or more green lights will come on. If there is difficulty operating the handle, look for a little lock that has to be moved out of the way, or try pulling the handle prior to lowering it. If no handle or gear lights can be found, the aircraft is presumably a fixed-gear type. But you don't need to worry about any of this; the controller will know, and will remind you what to do if anything needs to be done.

4. You are too low. Add more power at once!

5. You are too low. Add more power at once!

6. The controller will tell you.

7. Reduce the power slightly, while preventing the nose from dropping by gently pulling back on the yoke. The airplane will surely settle onto the runway. The important thing is to have it settle in a level or slightly nose-high attitude. Landing nose-down, nosewheel first, could have serious consequences.

8. With an instructor. What you can learn in a few hours will have lasting value to you. It is something like learning to ride a bicycle or to swim—once your brain and eyes and nerves and muscles have been there once, they won't ever forget the feel of it.